# *Seshat Anthology*
# *Volume 2*

# Seshat Anthology

## Volume 2

Austin Albert Mardon, Daivat Bhavsar, Katie Turner, Amy Zhao

Antarctic Institute of Canada

## Featured Article Themes

COVID-19 Communities

COVID-19 Mental Health and Activities

Space Exploration

Canadian Public Health Perspectives

## Financial Acknowledgment:

The publication of the Seshat Volume 2 was financially supported by the Antarctic Institute of Canada (AIC) and TakingItGlobal Charities.

**GM** PRESS

### *Editors*

Daivat Bhavsar

Katie Turner

Catherine Mardon


### *Graphic Designer*

Amy Zhao

### *Typeset*

Clare Dalton

# *List of Contributing Authors*

A.T. Ness

Alyssa Wu

Ananda Majumdar

Andy Kim

Austin Mardon

Catherine Mardon

Chitrini Tandon

Daivat Bhavsar

Daniel Polo

Dollyann Santhosh

Gordon Zhou

Isaac Oboh

James Fisher

Jasrita Singh

Jilene Malbeuf

John Christy Johnson

Katie Turner

Lina Lombo

Louis Park

Lucas Nowosiad

Muntaha Marjia

Peter Anto Johnson

Riley Witiw

Svetozar Zirnov

First Printing: 2021

Cover Design by Amy Zhao

Typeset by Clare Dalton

ISBN 978-1-77369-219-7

Golden Meteorite Press

103 11919 82 St NW

Edmonton, AB T5B 2W3

www.goldenmeteoritepress.com

# Contents

# Challenges Associated with COVID-19 Vaccinations

**Author**: Cheng'En Xi
**Affiliations:** Antarctic Institute of Canada
**Email:** aamardon@yahoo.ca

Much of Canada and Europe are now in the second wave of the COVID-19 pandemic. The pandemic that hit Canada in full force in March 2020 has dominated our lives for the past 8 months. From masks to closed businesses, from travel restrictions and online schools, it can be safe to say that everyone is tired of this pandemic.

Ever since the pandemic began, much of the hope of ending COVID-19 has been placed on the development of safe and effective vaccinations. Despite initial questions surrounding its feasibility, the world is now ever closer to the goal. Canada has also placed its bets, as the federal government has reached advanced purchasing agreements (APA) with many world-leading vaccine developers. These developers include companies like Moderna, AstraZeneca, Johnson & Johnson, Pfizer, with a combined dose of 262 million.

On November 9th, 2020, Pfizer made headlines announcing that their vaccine has demonstrated a 90% efficacy. This is no doubt an extraordinary achievement, given that not all vaccines are this efficacious. For instance, the seasonal flu vaccine can only reduce the risk of contracting the flu by 40-60%[1]. The announcement has led Prime Minister Justin Trudeau to use the phrase "light at the end of the tunnel"[2]. While optimism is certainly welcomed as the pandemic approaches its 1-year mark, one still needs to be cautious about what this announcement means. This optimism was displayed by US President-Elect Joe Biden when he said "It's clear that this vaccine, even if approved, will not be widely available for many months yet to come. The challenge before us right now is still immense and growing"[3].

Once countries approve the vaccine, it has to be produced and distributed in large numbers. Production is going to be a major challenge. With this pandemic reaching every corner of the globe, and many vaccines requiring two doses, there will need to be billions of doses of vaccines produced for COVID-19. This kind of vaccine production is unprecedented. For reference, the global seasonal flu vaccine production in 2015 is 1.467 billion doses, with the majority of manufacturing capacity being from higher-income countries[4]. Not to mention the fact that vaccines like the flu vaccine have pre-existing production and distribution networks, while much of the facilities for a COVID-19 vaccine are still being constructed. A major bottleneck in vaccine production is the bottles themselves[5]. According to a Reuters report in June 2020, vaccine makers around the world are scrambling to secure enough glass vials for their vaccines, with orders being in the hundreds of millions[5]. These vials may seem trivial, but they are absolutely necessary for the vaccine storage and distribution process. Production is going to be especially difficult when it is expected that more than 5.6 billion people globally need to be vaccinated to reach herd immunity, which can potentially mean at least 11 billion vaccine doses[6].

Even if the production problem is solved, there is still distribution. A major obstacle any vaccine needs to overcome when it is approved is the cold chain. Much of the vaccines in development are mRNA vaccines, such as the ones by Pfizer and Moderna, which are highly unstable, and require temperatures as low as -70 degrees[7]. To combat this issue, Pfizer has produced specialized containers that use dry ice to maintain the temperature. However, the containers have a set lifespan of 10-15 days and contain 1,000-5,000 doses of vaccines[7]. So realistically, accounting for transportation time, there are not that many days for vaccinations to take place before the vaccines in the container begin to expire if the dry ice cannot be properly replenished. The same challenge would apply to all other mRNA vaccines. The low temperatures required by some of these vaccines means that injection sites need to have either the infrastructure to support the long term ultra-cold storage of these vaccines or have the ability to vaccinate large numbers of people in a short period of time while having dry ice on hand for replenishment. Furthermore, there is the issue of global distribution. If every country sought to secure as many vaccines as possible for their own people, this could lead to a bidding war between countries, leaving poorer countries out in the cold. This can lead to a similar situation as securing PPEs early in the pandemic, where severe supply and demand imbalance made the PPE industry into "the Wild West"[8]. This can add more chaos and strain to the already complicated vaccine distribution network. The WHO has since started a global vaccination initiative called COVAX, aimed at ensuring developing countries would also have access to these life-saving vaccines by coordinating global facilities[9]. However, much of the agreement is non-binding and the US has yet to sign on[10].

Finally, there is the question of how many people are actually willing to receive the vaccine. The COVID-19 pandemic has also brought on a pandemic of misinformation and conspiracy theories. Misinformation such as 5G, Plandemic, and Bill Gates microchip conspiracies have managed to reach audiences in the millions and contribute significantly to the anti-lockdown protests we see today[11]. The age of social media is bringing new challenges to vaccination as misinformation can spread much faster and reach a wider audience. This will undoubtedly affect the public's enthusiasm for getting a COVID-19 vaccine. Polls show that more than half of Canadians believe that vaccines should be voluntary[12] and that roughly a third of Canadians are hesitant about getting a COVID-19 vaccine and would prefer to wait and see[13]. The same polls also show that enthusiasm for a vaccine has been dropping in Canada. Their hesitancy is not completely unreasonable, as vaccine developers such as Johnson & Johnson and AstraZeneca all underwent trial pauses due to unexplained adverse events[14,15]. However, reluctance to be vaccinated can prolong the pandemic, even if a vaccine is successfully produced and distributed. Herd immunity requires that a high percentage of the population needs to be immune, which, depending on the virus's transmissibility and the vaccine's efficacy, can range anywhere from 60-90%[16]. So, vaccine hesitancy is another hurdle that needs to be overcome and is arguably the most unpredictable factor in the path to vaccination.

Despite the optimism surrounding vaccine development, realistically, it will still be many months until the general public in Canada will be able to get a COVID-19 vaccine. As cases are surging in much of Canada, patience and caution are still absolutely necessary in this protracted war against an invisible enemy. The road to the end of this pandemic will be a marathon, not a sprint, and the successful development of a vaccine is still just the beginning. Many challenges will still need to be overcome until normal life and safely resume. Until then, wear a mask, stay socially distant, and do your own part in helping to avoid a "dark winter"[3].

## References

Vaccine Effectiveness: How Well Do the Flu Vaccines Work? | CDC [Internet]. 2020 [cited 2020 Nov 13]. Available from: https://www.cdc.gov/flu/vaccines-work/vaccineeffect.htm

Trudeau says Pfizer vaccine could be "light at the end of the tunnel," but hurdles remain [Internet]. nationalpost. [cited 2020 Nov 13]. Available from: https://nationalpost.com/news/politics/trudeau-says-pfizer-vaccine-could-be-light-at-the-end- of-the-tunnel-but-hurdles-remain

Sabga P. Biden urges vigilance as Pfizer COVID-19 vaccine spurs hopes [Internet]. [cited 2020 Nov 13]. Available from:

https://www.aljazeera.com/economy/2020/11/9/biden-urges-vigilance-as-pfizer-covid-19-va ccine-spurs-hopes

McLean KA, Goldin S, Nannei C, Sparrow E, Torelli G. The 2015 global production capacity of seasonal and pandemic influenza vaccine. Vaccine. 2016 Oct 26;34(45):5410–3.

Blamont LB Matthias. Exclusive: Bottlenecks? Glass vial makers prepare for COVID-19 vaccine. Reuters [Internet]. 2020 Jun 12 [cited 2020 Nov 14]; Available from: https://www.reuters.com/article/us-health-coronavirus-schott-exclusive-idUSKBN23J0SN

Rowland C. A race is on to make enough small glass vials to deliver coronavirus vaccine around the world. Washington Post [Internet]. [cited 2020 Nov 14]; Available from: https://www.washingtonpost.com/business/2020/07/13/coronavirus-vaccine-corning-glass/

Pfizer readies "Herculean effort" to distribute coronavirus vaccine [Internet]. NBC News. [cited 2020 Nov 14]. Available from:

https://www.nbcnews.com/business/business-news/pfizer-readies-herculean-effort-distribute-coronavirus-vaccine-n1247147

Ottawa and the provinces are navigating a 'Wild West' in the medical supply market. [cited 2020 Nov 14]; Available from:

https://www.theglobeandmail.com/politics/article-ottawa-and-the-provinces-are-navigating-a-wild-west-in-the-medical/

172 countries and multiple candidate vaccines engaged in COVID-19 vaccine Global Access Facility [Internet]. [cited 2020 Nov 14]. Available from: https://www.who.int/news/item/24-08-2020-172-countries-and-multiple-candidate-vaccines- engaged-in-covid-19-vaccine-global-access-facility

Rauhala E, Abutaleb Y. U.S. says it won't join WHO-linked effort to develop, distribute coronavirus vaccine. Washington Post [Internet]. [cited 2020 Nov 14]; Available from: https://www.washingtonpost.com/world/coronavirus-vaccine-trump/2020/09/01/b44b42be-e965-11ea-bf44-0d31c85838a5_story.html

Covid: London anti-lockdown protest leads to 190 arrests. BBC News [Internet]. 2020 Nov 6 [cited 2020 Nov 14]; Available from:

https://www.bbc.com/news/uk-england-london-54842605

Canadians divided over mandatory COVID-19 vaccines, priority inoculations - CityNews Toronto [Internet]. [cited 2020 Nov

14]. Available from: https://toronto.citynews.ca/2020/10/13/cana-da-covid19-vaccine-poll/

Jung A. New survey finds more Canadians are hesitant about getting a vaccine against COVID-19 [Internet]. British Columbia. 2020 [cited 2020 Nov 14]. Available from: https://bc.ctvnews.ca/new-sur-vey-finds-more-canadians-are-hesitant-about-getting-a-vaccine-against-covid-19-1.5131271

Burger L. AstraZeneca pauses coronavirus vaccine trial as par-ticipant illness investigated. Reuters [Internet]. 2020 Sep 9 [cited 2020 Oct 22]; Available from: https://www.reuters.com/article/us-health-coro-navirus-astrazeneca-idUSKBN26017L

CNN MF. Johnson & Johnson pauses Covid-19 vaccine trial after "unexplained illness" [Internet]. CNN. [cited 2020 Nov 14]. Avail-able from: https://www.cnn.com/2020/10/12/health/johnson-coronavi-rus-vaccine-pause-bn/index.html

Anderson RM, Vegvari C, Truscott J, Collyer BS. Challenges in creating herd immunity to SARS-CoV-2 infection by mass vaccination. The Lancet [Internet]. 2020 Nov 4 [cited 2020 Nov 14];0(0). Available from:

https://www.thelancet.com/journals/lancet/article/PIIS0140-6736(20)32318-7/abstract

**Author Biography:**

Cheng'En Xi is an undergraduate student at McMaster
University passionate about the healthcare challenges magnified by
COVID-19 and serves as an independent Article Writer Volunteer under
the Antarctic Institute of Canada.

# COVID-19 Tent Cities
# in Western Canada

Katie Turner*, Marika Forsythe, Daivat Bhavsar, Jasrita Singh, Austin Mardon

## Abstract

This article analyzes two tent cities, Camp Pekiwewin and the Kennedy Trudeau Camp, which were established in Edmonton and Vancouver respectively during COVID-19. COVID-19 influenced the development of these groups due to the economic effects of the pandemic, the added vulnerability of homeless people during the pandemic, and the lack of support these people received from their governments. This article discusses the outcomes of the camps and their potential futures after COVID-19.

## Keywords:

COVID-19; Pandemic; Homeless; Tent cities; Western Canada; Vulnerable Populations

## Introduction

The COVID-19 pandemic influenced the formation of large tent cities - temporary and informal housing groups made of tents and various other makeshift shelters - that were organized by nonprofits and volunteers. They were a place for homeless people to live, but they also made a list of changes they wanted the city to make to help those people. Two of the many camps emerged in large western Canadian cities - Vancouver and Edmonton. The Edmonton camp- Pekiwewin - was originally a sacred indigenous gathering place, located a few blocks from downtown in a baseball stadium parking lot. The camp was erected on July 24, 2020 by various frontline workers and Indigenous community members. Initiated with about a dozen tents, the camp grew to around 200 over the course of a couple of months. The Kennedy Trudeau (KT) camp in Vancouver was also near downtown and was named after Prime Minister Justin Trudeau and Vancouver Mayor Kennedy Stuart. The camp began in June and at its peak had nearly 400 residents. This article will examine how COVID-19 has influenced the development of these camps, the demands of their inhabitants, and the effects from these camps at present and after the pandemic. Although this article focuses on two specific western Canadian camps, the information we gain from analyzing them could help other cities and countries learn how to better help this vulnerable population in similar situations.

## How COVID-19 Influenced Tent Cities

Tent cities, and specifically-organized tent cities, are not a new phenomenon.[1,2] Tent cities have existed in the United States since the Civil War, although they were originally home to seasonal workers, during the Great Depression more unemployed people began to live in these areas, and in the 1990s many of these camps became more permanent.[3] There are implications of the COVID-19 pandemic that have influenced the development of some of these camps. COVID-19 and the associated struggles have exacerbated issues that were already present prior to the pandemic such as the economic downturn, the added vulnerability of homeless people, and the lack of government support.

# Economic Downturn

An increase in homelessness is often correlated with an economic crisis. For example, between 2007 and 2008 the USA had a huge economic downturn and national data showed that homelessness more than doubled.[2] The inability of people to pay their bills and afford basic living expenses can mean more evictions and loss of stable living, thereby contributing to the increased rates of homelessness. The public health restrictions due to COVID-19 have severely affected Canada's economy. There have been layoffs, business closures, and with schools being closed many parents were forced to stay home. A survey done by Statistics Canada showed that over half of all businesses saw a revenue decline of 20% in March.[4] The same study found that 38% of businesses reduced staff hours while 40% reported they laid off staff.[4] These declines affect the business owners and their employees, meaning more people struggling to have make a sustainable income to afford their basic needs such as a permanent place to live.

# Added Vulnerability of the Homeless Population

Canadians have become accustomed to the advice given by medical officials, mainly washing your hands, social distancing whenever possible, and staying home when sick. Although this advice may seem easily followed, in practice, it poses great challenges for homeless people, especially during the winter season and for those living in shelters.[5] Homeless people do not always have access to running water, shelter living makes it near impossible to socially distance, and staying home when feeling sick is not an option when they do not have their own home to go to. The CDC recommended that shelters have people sleep 6 feet apart and that they provide individual rooms for those exhibiting COVID-19 symptoms or confirmed cases.[6] This is very difficult for many shelters that are already overcrowded. In

2019, the Canadian census revealed that there were 22,190 people living in shelters in Canada but Statistics Canada declared in 2020 there were only 15,400 emergency shelter beds available.[7,8] This disproportionate number of beds available to those in need has only worsened due to the pandemic health recommendations, and with no where else to go, people are forced to live in these conditions.

On top of the inability to abide by health recommendations, homeless people are further at risk due to pre-existing medical conditions. Homeless populations have a higher rate of medical illness, psychopathology, and substance abuse.[9] Pre-existing health issues such as hypertension, diabetes, and coronary heart disease are related to fatal cases of COVID-19.[10] Due to increased exposure, overcrowding, substance abuse, and other co-morbid disorders, homeless people have an increased risk of tuberculosis, influenza, and pneumonia as well.[11] Tuberculosis is also a risk factor for more severe cases of COVID-19.[12] Air ventilation and air flow are an issue in many shelters, particularly those not originally built to house so many people in close proximity. This, along with overcrowding, can be a contributing factor in the increased transmission of infectious diseases like tuberculosis.[13] The aforementioned medical conditions and risk factors that homeless people face on a regular basis significantly increase the likelihood of severe outcomes from COVID-19. With disease running rampant in homeless shelters, some people may find the better option to be living outside where there is more room and proper air flow lowering the risk of contracting COVID-19. This results in the more people living in tents or other makeshift shelters, which develop into larger tent cities.

## The Lack of Government Support

Although the homeless population is at high risk for contracting COVID-19, many of these individuals feel local and provincial governments have not made an adequate effort to help. Pekiwewin's website claims that during the pandemic, overnight shelters, services, and public washrooms have been unavailable and the police committed acts of violence against homeless people throughout.[14] There were no news articles that reported these acts taking place at Pekiwewin, however police calls were up 113% in the area since the camp was erected indicating a large police presence in the area.[15]

The KT camp was formed under similar circumstances. Prior to KT, a large tent city in Oppenheimer Park was shut down by police. Many homeless people were then kicked out of the next camp they went to and thus the KT tent city was formed. As police push people into smaller spaces with more rules, and shelters remain underfunded and unable to keep up with the population, these larger scale long term tent cities continue to pop up.[16] The principle of utility is about creating the "greatest good for the greatest amount of people", but this often overlooks those who are marginalized, including the homeless population.[5]

## Their Demands

The request of camp Pekiwewin to the city of Edmonton were the following: 1) an end to the disruption of homeless camps and homeless people's possessions; 2) review of racist bylaws; 3) more transitional support services; 4) emergency response fund for front line workers and communities with lived experience of homelessness; 5) free transit; and, 6) for the Pekiwewin grounds to remain a ceremonial gathering site.[14] KT gave similar demands to their city including permanent housing, ending displacement, and reclaiming indigenous land.[17]

Both camps were largely being run by people who experienced homelessness and/or have worked in a large capacity with homeless people. Most public health policies are not created by people who have these experiences and therefore may not understand all the issues faced by these populations.[18] These camps are a way for homeless people to have their voices heard and give them a sense of autonomy.

## Government Reactions & Future Possibilities

The Mayor of Edmonton, Don Iveson, stated in August 2020 that they were creating a 10-week plan to end homelessness.[19] While this deadline has long since passed, there has still been little done other than creating another emergency shelter. None of Pekiwewin's demands have been met. Edmonton now has two large scale emergency shelters at the Convention Centre and a stadium, one of which currently has a COVID-19 outbreak and therefore reduced their capacity to half.[20] This does not fix the problem at hand, as these are only temporary shelters and it re-introduces the issue of overcrowding and increased risk during the pandemic. The city of Vancouver has taken a larger step towards helping by allocating $30 million to housing homeless people during the pandemic.[21] Again it fails to be a long term solution or address the camps demands, but it does reduce the risk of COVID-19 spreading rampant in this vulnerable population.

Pekiwewin and KT are not the first of these organized homeless camps. Some cities, particularly in the United States, have allowed for sanctioned camps that are semi-autonomous. However, they often are given very strict rules and displaced far away from city centers.[15] Being far away from downtown means it is more difficult for people to access resources that are largely accumulated in a city's center. Both Pekiwewin and KT were in very public places close to downtown, which may have led to issues as city officials and residents tried to push them further out of sight.

Western Canadian tent cities have another factor involved as well, dealing with the harsh reality of Canadian winters. Camp Pekewewin did not continue long after snowfall. Pekiwewin declared that their services would be stopped on November 6, 2020. Some people chose to continue living in the tents for a time, but Edmonton police began removing people and their belongings shortly after that date. As of December 15, 2020 camp KT continues to run although the city is trying to move residents to allocated temporary hotels.

## Conclusion

COVID-19 has created a perfect storm for community protests, as can be seen by the tent cities that developed in Edmonton and Vancouver and the demands being made. Local governments did not disband the camps immediately, but they did not support them or cave to the demands. Although similar camps have been legalized in some cities, the harsh weather of Western Canada poses particular difficulties. If not given more funding or more permanent housing solutions the camps may indeed return in the spring, requiring the local governments to finally make decisions regarding their management. Canada is not the only country facing similar challenges, tent cities have increasingly developed in other western countries during the COVID-19 pandemic.

COVID-19 has brought some of the most vulnerable peoples in our societies, often the ones most overlooked, to the forefront, and governments will need to choose what to do with this.

### Acknowledgement

This article was published with funding from the Rising Youth Grant.

# References

Herring C, Lutz M. The Roots and Implications of the USA's Homeless Tent Cities. Analysis of Urban Change, Theory, Action. 2015; 19 (5).

Loftus-Farren Z. Tent Cities: An Interim Solution to Homelessness and Affordable Housing Shortages in the United States. California Law Review 2011; 99: 1037-1082.

Herring C. Tent City, America. Places 2015.

Statistics Canada. Canadian Survey on Business Conditions: Impact of COVID-19 on businesses in Canada, March 2020.

https://www150.statcan.gc.ca/n1/daily-quotidien/200429/dq200429a-eng.htm. 29 April, 2020.

Silva,D, Smith M. Social Distancing, Social Justice, and Risk During the COVID-19 Pandemic. Canadian Journal of Public Health 2020; 111: 459-461.

CDC. Interim Guidance for Homeless Service Providers to Plan and Respond to Coronavirus Disease 2019 (COVID-19).

https://www.cdc.gov/coronavirus/2019-ncov/community/homeless-shelters/plan-prepare-respond. tml. 21 April, 2020,

Statistics Canada. The Population living in shelters who are they?

https://www150.statcan.gc.ca/n1/daily-quotidien/190415/dq190415a-eng.htm. 15 April, 2019.

Stats Canada. Canada announces support to those experiencing homelessness and women fleeing gender-based violence during the coronavirus disease (COVID-19) pandemic.

https://www.canada.ca/en/employment-social-development/

news/2020/04/canada-announces-sup

port-to-those-experiencing-homelessness-and-women-fleeing-gender-based-violence-during-the-c onavirus-disease-covid-19-pandemic.html. 4 April, 2020.

Schanzer B, Dominquez B, Shrout PE, Caton CLM. Homelessness, health status, and health care

use. Am J Public Health. 2007;97(3):464-469.

Chen R, Liang W, Jiang M, Guan W, Zhan C, Wang T, et al. Risk Factors of Fatal Outcome in Hospitalized Subjects with Coronavirus Disease 2019 from a Nationwide Analysis in China. Chest. 2020; 158(1): 97-105

Wrezel O. Respiratory Infections in the Homeless. UWO Medical Journal. 2009; 78(2):61-65.

Chen Y, Wang Y, Fleming J, Yu Y, Gu Y, Liu C, et al. Active or Latent Tuberculosis Increases Susceptibility to Covid-19 and Disease Severity. Medrxiv. 2020. 1-15.

Moffa M, Cronk R, Fejfar D, Dancausse S, Acosta Padilla L, Bartram J. A Systematic Scoping

Review of Environmental Health Conditions and Hygiene Behaviors in Homeless Shelters. International Journal of Hygiene and Environmental Health. 2019. 335-346.

Pekiwewin Demands. https://pekiwewin.com/support-our-demands

Police Calls up 113 per cent in Rossdale Near Camp Pekiwewin Over Last Year: Chief McFee.

Edmonton Journal.

https://edmontonjournal.com/news/local-news/police-calls-up-

113-per-cent-in-rossdale-near-cam ppekiwewin-over-last-year-chief-mc-fee. 8 October, 2020.

Speer J. The Rise of the tent ward: Homeless camps in the era of mass incarceration. Political Geography. 2018; 62: 160-169.

As Vancouver Tent City Expands, Some Neighbours Voice Support for Campers' Demands by CBC.

https://www.cbc.ca/news/canada/british-columbia/neighbours-new-tent-city-are-supportive-of-camper-demands-1.5636487 4 July, 2020.

Silva DS, Smith MJ, & Upshur REG. (2013). Disadvantaging the disadvantaged: when public health policies and practices negatively affect marginalized populations. Canadian Journal of Public Health, 104(5), e410–e412.

Mayor Sets 10-week Due Date for Homeless Strategy by CTV.

https://edmonton.ctvnews.ca/mayor-sets-10-week-due-date-for-homeless-strategy-1.5084041 28 August, 2020.

Edmonton Convention Centre Shelter COVD-19 Outbreak Grows to 42 Cases, On-Site Testing to be Offered for Close Contacts by Edmonton Journal.

Edmonton-convention-centre-outbreak-update. 1 December, 2020.

COVID-19: City of Vancouver Passes $30 Million Homeless Action Plan by Vancouver Sun.

https://vancouversun.com/news/covid-19-city-of-vancouver-passes-30-million-homeless-acti on-plan. 10 October, 2020.

**Biographies**

Katie Turner is a writer, peer support worker, and mental health advocate. She currently holds a BA in psychology and sociology from the University of Alberta. Katie recently co-authored a book titled 'Alphabet Soup' which discusses various mental health disorders and the experience of living with them.

Marika Forsythe is a recent graduate of Saba University Medical School. She is hoping to match into a pathology residency this year.

Daivat Bhavsar is an undergraduate student with a background in Biochemistry at McMaster University.

Jasrita Singh is an undergraduate student with a background in Biochemistry, Biomedical Discovery and Commercialization at McMaster University.

Austin Albert Mardon, CM Ph.D. is an author, community leader, and advocate for mental health. He is an assistant adjunct professor at the John Dossetor Health Ethics Centre at the University of Alberta. In the mid 80's, he founded and today still directs the Antarctic Institute of Canada, a non-profit entity based in Edmonton, Alberta. He is also an Order of Canada member and Fellow of the Royal Society of Canada.

# Evaluation of the Exclusion of Dental Services from Essential Medical Services during COVID-19

Louis S. Park[*1,2], John Johnson [1,3], Peter Johnson [1,3], Jilene Malbeuf [1,3], Jasrita Singh [1,2], Austin A. Mardon, CM Ph.D.[1,3]

[1] Antarctic Institute of Canada, Edmonton AB, Canada

[2] McMaster University, Faculty of Health Sciences, Hamilton ON, Canada

[3] University of Alberta, Edmonton AB, Canada

## Introduction

The COVID-19 pandemic placed many non-essential services on hold. However, the boundaries between essential and non-essential services are unclear in the medical sector, particularly in regards to dentistry. Essential services are daily services essential to preserving life, health, public safety and basic societal functioning. Hence, amid the COVID-19 pandemic, hospitals are triaging and treating patients normally, accepting a wide range of patients with varying degrees of emergency and illnesses that pose a risk on an individual's health and safety.

However, bearing the same risk of transmission and equally preserving public health, most dental procedures have been suspended, and resorted to telemedicine. Dentists are not well protected in their work environments and people who are facing minor dental illnesses are left untreated indefinitely, increasing their risk of severe, dental illnesses in the future.

We discuss the guidelines of essential medical services set out in 2020 during the COVID-19 pandemic and implications and specific challenges that arise from the closure of 'non-essential' dental treatments.

## Which dental procedures are considered essential?

The Royal College of Dental Surgeons of Ontario (RCDSO) has only declared dental emergency procedures as essential, which constitutes trauma, infection, prolonged bleeding, and severe pain. [1] All other regular urgent procedures that would maintain a patient's oral health were halted indefinitely. Also, the RCDSO has further restricted dentistry through the use of teledentistry; a virtual provision of professional dental advice via communication technologies [2]—one not being readily used or recommended in hospitals. Some Canadians lack communication devices and most dental offices are not technologically prepared to offer teledentistry. This limits the number of patients they can manage and leads to the closure of dental clinics, placing personal financial consequences on dentists as well. In fact, about 78.7% of dental emergencies presented at hospitals are semi- to non-urgent and are diverted to local dental offices, leaving these patients without guaranteed treatment. [3] This would further affect COVID-19 response as it would be difficult to relieve emergency rooms, maintain low cost and high quality of care during a pandemic.

Furthermore, oral infections are known to affect the pathogenesis of systemic diseases including cardiovascular disease, pneumonia, and diabetes. [4] However, with indefinite halt in place, many dental patients are unable to prevent or detect time-sensitive dental illnesses like tooth decay, gum disease, oral cancer or vitamin deficiencies that would normally have been treated. This is especially problematic when considering the large population of frail and elderly individuals in Canada who are significantly associated with poor oral health. [4] Hence, the exclusion of dental care services poses a threat on the oral health of Canadians, thereby increasing risk of possible systematic diseases, poor quality of life, and if untreated, time-sensitive dental problems.

## What special COVID-19 preventative challenges arise in dental clinics?

COVID-19 is known to spread via bodily fluid particles such as mucus and saliva, substances dentists are connected with during dental procedures. Additional challenges arise from the nature of the procedures in which the clinician must be in great proximity, face-to-face with the patient, making it impossible for dentists to comply with the social distancing measures. In addition, with the scientific brief from the World Health Organization (WHO), which has suggested that COVID-19 may be capable of airborne transmission, [5] dentists are at an even higher risk for COVID-19. This may substantiate and justify the exclusion of dentists who are most vulnerable and at a greater risk of transmission of COVID-19 from essential medical services for their protection.

Nonetheless, while these safety measures and the closure of dental services maintain safety and protection from the coronavirus, this poses significant personal financial consequences on private offices and employees. [6]

# Personal Protective Equipment supply and demand

The College of Dental Hygienists of Ontario and RCDSO has established high-demanding, preventive guidelines and protocols. These protocols mandate the use of N95 masks, gowns, and other personal protective equipment (PPE) such as face shields and gloves, which demand a higher supply. As COVID-19 cases are increasing, it is crucial to allocate PPE to those in the hospitals interacting with active COVID-19 patients, especially during a period of shortage.

With such high demands in place with an inadequate supply, the "essential" boundary becomes apparent. Clinicians interacting with active patients should rightfully be considered more "essential" and be given priority in regards to PPE.

However, according to the Priority setting of PPE released by the Ontario Provincial government, the Secondary Allocation Principle relies on first-come-first-serve or lottery, [7] which is not risk level- nor demand-focused. These principles limit dentists' allocation for PPE, threaten their safety, and it raises a concern of whether every dental office would even be able to provide emergency care.

# Conclusion

While essential medical services such as emergency rooms directly treating COVID-19 patients should be prioritized, public health control measures pose challenges to dental services and increase the risk of poor oral health of Canadians amid indefinite lockdown measures. Safety protocols have been implemented in Canada to prioritize essential COVID-19-related healthcare, which caused a redeployment and reduction of dental clinicians. Essential medical care should be expanded to include more urgent dental procedures, not limited to virtual appointments nor to symptoms that classify as emergencies. The diverse implications must also be re-considered to ensure the well-being and health of citizens and clinicians, as well as an equitable and need-based distribution of PPE. The COVID-19 pandemic raises the question of whether the current measures on halting dental procedures must be re-evaluated for future pandemics.

## Acknowledgement

We thank and acknowledge the financial support of the Canada Service Corps, TakingITGlobal, and the Government of Canada.

## References

Definitions of emergency, urgent and non-essential care. *Royal College of Dental Surgeons of Ontario*. May 2020. PDF.

https://az184419.vo.msecnd.net/rcdso/pdf/standards-of-practice/ RCDSO_COVID19_Definition s.pdf. Accessed August 4, 2020.

Jampani ND, Nutalapati R, Dontula BS, Boyapati R. Applications of teledentistry: A literature review and update. *J Int Soc Prev Community Dent*. 2011;1(2):37-44. doi:10.4103/2231-0762.97695

Wall T, Nasseh K, Vujicic M. Majority of dental-related emergency department visits lack urgency and can be diverted to dental offices. Health Policy Institute Research Brief. American

Dental Association. August 2014. http://www.ada.org/~/media/ADA/
Science%20and%20Research/HPI/Files/HPIBrief_0814_1.as hx


Li X, Kolltveit KM, Tronstad L, Olsen I. Systemic diseases
caused by oral infection. *Clin Microbiol Rev*. 2000;13(4):547-558.
doi:10.1128/cmr.13.4.547-558.2000


Transmission of SARS-CoV-2: implications for infection
prevention precautions. World Health Organization.

https://www.who.int/news-room/commentaries/detail/
transmission-of-sars-cov-2-implications-f or-infection-prevention-
precautions. Published July 9, 2020. Accessed August 4, 2020.


Coulthard P. Dentistry and coronavirus (COVID-19) - moral
decision-making. *British Dental Journal*. 2020;228(7):503-505.
doi:10.1038/s41415-020-1482-1


Ontario Health. Priority Setting of Personal Protective
Equipment – Within Health Care Institutions and Community Support
Services. Published March 25, 2020. Accessed August 6, 2020. https://
www.wrh.on.ca/uploads/Coronavirus/Ethics_Table_Policy_Brief_3_
PPE_Within_Health

_Care_Institutions_Community_Support_Services.pdf

# Biographies

## Louis S. Park

**John Christy Johnson, BSc(Hons).** is a writer, biomedical engineer, and mental health advocate with a strong passion for accessibility and community service. He currently BSc (Hons) degree from the University of Alberta and has been recognized for his notable community and research achievements, having been named one of Alberta's Top 30 Under 30 in 2019.

**Peter Anto Johnson, MSc, BSc(Hons).** is a writer, medical scientist, and mental health advocate with a strong passion for healthcare. He currently holds an MSc in medical sciences and a BSc (Hons) degree from the University of Alberta and was recognized for his notable community and research achievements, being named one of Alberta's Top 30 Under 30 in 2019.

## Jilene Malbeuf

**Jasrita Singh** is an undergraduate student with a background in Biochemistry, Biomedical Discovery and Commercialization at McMaster University.

**Austin Albert Mardon, CM Ph.D.** is an author, community leader, and advocate for mental health. He is an assistant adjunct professor at the John Dossetor Health Ethics Centre at the University of Alberta. In the mid 80's, he founded and today still directs the Antarctic Institute of Canada, a non-profit entity based in Edmonton, Alberta. He is also an Order of Canada member and Fellow of the Royal Society of Canada.

# How the pandemic revealed a public washroom crisis

**Author**: Cheng'En Xi

**Affiliations:** Antarctic Institute of Canada

**Email:** aamardon@yahoo.ca

The COVID-19 pandemic has exposed many flaws in how our society normally operates. Systems that previously operate adequately at or near their full capacity are now overflowing and exposing its underlying issues that were hidden before the pandemic. One such system is how commuters access public washrooms. Unlike issues such as unemployment and business closures, this issue does not get the attention it deserves. While it does not threaten livelihoods, it does present a common inconvenience for many.

Even before the pandemic, public washroom access remains a problem for many. The Toronto subway system has more than one million riders[1] on an average workday before the pandemic, and for a total of 75 operational stations, just a meagre 11 of them have public washroom access[2]. That means if you are a commuter who has to travel across the city, while having the misfortune of entering and exiting at stations without washrooms, you better be mindful of your hydrations. Before the pandemic, as inconvenient as this is, people often deal with it by going to the nearest Tim Horton's, or other similar establishments to use their washrooms. However, now this is no longer doable for many. Many fast-food restaurants no longer allow washrooms for public use, or at best, require someone to ask for a key at the counter. While this gesture is understandable due to the fear of spreading the virus, it presents a major inconvenience for many commuters. This also negatively impacts those whose jobs require them to be constantly on the move, such as driving instructors, taxi and Uber drivers, and truck drivers. In fact, the problem was so pronounced that back in late March 2020, Ontario Premier Doug Ford publicly pleaded with business owners for them to keep washrooms open for truck drivers[3].

In the era of COVID-19, the operation of anything cannot go ahead unless the transmission of the virus is considered. So, how safe are public washrooms? Unfortunately, there is no clear-cut answer. Being an indoor space, a public washroom will come with the same hazards as other indoor activities and is dependent on ventilation, duration of stay, and how many other people are present. Recent studies done at hospitals at Wuhan found that washrooms contained high concentrations of aerosol, which is a vector of transmission. However, it was a temporary, single-toilet room with no ventilation. Another study found that they were unable to isolate virus samples from the stool of COVID-19 patients, and existing viral fragments were not infectious. Furthermore, much of the aerosol is produced by toilet flushes, and if the aerosol contained any virus, the person is likely already infected since he is the source of the stool. In addition, other more socially acceptable indoor events, and activities, such as weddings, bars, and religious services, can be riskier due to the long exposure time, the number of densely packed people, and the relaxed atmosphere that makes people converse and forget social distancing[4]. So, despite the additional ick factor, public washrooms may not be any more dangerous than other indoor activities, while being absolutely essential.

The good news about this issue is that it is finally getting some attention at the governmental level. In November 2020, the city of Toronto has announced that the city is opening 79 new "winter washrooms" in parks and other outdoor recreational facilities[5]. This is definitely going to alleviate some of the problems, especially since a new round of lockdown is announced for the holiday season. However, this decision is far from a permanent solution. Once the pandemic ends, it is likely that these washrooms will continue to be closed for future winters, and the problem will not be permanently resolved. Furthermore, as transit expansion plans are being considered for the GTA region, including subway extensions into Scarborough and Richmond Hill[6], this is a great opportunity to increase washroom access by having one at each of the new stations, not just the terminal stations. This is especially important as it expands into less densely populated areas, where finding public washrooms outside of the station would be even harder than it is downtown. Now that the pandemic has revealed the shortcomings of the existing public washroom system, it presents a great opportunity for change and improvement, and it is an opportunity that needs to be taken.

**References:**

2019-Q1-Ridership-APTA-1.pdf [Internet]. [cited 2020 Dec 21]. Available from: https://www.apta.com/wp-content/up-loads/2019-Q1-Ridership-APTA-1.pdf

The closest public washroom to every Toronto subway station (MAP) | Urbanized [Internet]. [cited 2020 Dec 21]. Available from:

https://dailyhive.com/toronto/toronto-public-wash-room-map-2017

'Have A Heart' And Keep Public Washrooms Open For Truck Drivers: Ford | HuffPost Canada [Internet]. [cited 2020 Dec 21]. Available from: https://www.huffingtonpost.ca/entry/doug-ford-truck-drivers_ca_5e83b1b2c5b65dd0c5d5ad 2c

Calechman S. How risky is using a public bathroom during the pandemic? [Internet]. Harvard Health Blog. 2020 [cited 2020 Dec 21]. Available from: https://www.health.harvard.edu/blog/how-risky-is-us-ing-a-public-bathroom-during-the-pand emic-2020071420556

Toronto is opening up 79 new washrooms for public use this winter [Internet]. [cited 2020 Dec 23]. Available from:

https://www.blogto.com/city/2020/11/toronto-opening-79-new-public-washrooms-winter/

Transit expansion in the Greater Toronto Area [Internet]. Ontario.ca. 2020 [cited 2020 Dec 23]. Available from: https://www.ontario.ca/page/transit-expansion-greater-toronto-area

**Author Biography:**

Cheng'En Xi is an undergraduate student at McMaster University passionate about the healthcare challenges magnified by COVID-19 and serves as an independent Article Writer Volunteer under the Antarctic Institute of Canada.

# The impact of the COVID-19 pandemic on mental health among adults: with a focus on depression data

**Authors:** Muntaha Marjia[1,2], Lucy Chen[1,3], Tian Jian Gao[1,4], and Dr. Austin Mardon[1,5]

**Author Affiliations:**

[1]Antarctic Institute of Canada, #103 11919 82 St NW, Edmonton AB, T5B 2W4, Canada

[2]Biomedical Sciences, Faculty of Science, Ryerson University Toronto, Ontario

[3]Honours Specialization in Physiology, Medical Sciences,Faculty of Science and Schulich School of Medicine and Dentistry,Western University, London Ontario

[4]Medical Sciences, Faculty of Science and Schulich School of Medicine and Dentistry, Western University, London, Ontario

[5]Department of Psychiatry & John Dossetor Health Ethics Centre, University of Alberta, 2J2.00 WC Mackenzie Health Sciences Centre, 8440 112 St NW, Edmonton AB, T6G 2R7, Canada

As a result of the on-going COVID-19 pandemic, the year 2020 has been like no other. From becoming unemployed to being unable to visit loved ones, the repercussions of the pandemic has been felt by many. As a result of the negative changes caused by the COVID-19 pandemic, maintaining a healthy mental state can be challenging. In fact, numerous studies that focus on young adults (individuals aged 18-30 years old) have found an increase in issues associated with mental health and specific mental illnesses among the population.

Since the World Health Organization declared a global pandemic for SARS-CoV-2 on March 11, 2020, young adults have encountered many drastic changes to their daily lives. These changes include, but are not limited to: shifting to an online school environment, limited face-to-face interactions, social distancing, and reduced employment opportunities. Adjusting to these unfavourable circumstances may have excessively placed mental strain on young adults and consequently heightened the effects and/or the development of mental health problems[2]. An online study conducted by the Brigham and Women's Hospital located in Boston, Massachusetts surveyed 898 young adults living in the U.S. from April 13, 2020 to May 19, 2020[3]. Participant responses were conducted using a qualitative measure of various components of their mental state. For example, the study gathered data on depression amongst participants by using the Patient Health Questionnaire (PHQ) consisting of 8 items (PHQ-8). The PHQ is a credible assessment of the severity of depressive disorders amongst participants of large clinical studies[4]. Within the PHQ-8, individuals were asked to rank the recurrence of symptoms connected to depression during the two weeks period prior to their completion of the questionnaire. The survey was listed in the format, (0) representing "never" to (3) representing "almost every day" for the possible answers. The study found 43.3% of respondents identified with higher levels of depression, and this was represented by a PHQ-8 score of 10 or greater (the possible range of scores was 0-24)[3]. A similar study was conducted in the United Kingdom by researchers from the University of Nottingham, King's College London and the University of Auckland between April 3, 2020 and April 30, 2020. In this study, 3097 individuals of which 364 were ages 18-24 years old completed a survey relating to mental health[5]. Similar to the previous U.S. study mentioned, participants completed a Patient Health Questionnaire with 9 items (PHQ-9). The portion of respondents between the ages of 18-24 (i.e. young adults) reported a mean PHQ-9 score of 11.23. This value is significantly higher than the standard score for PHQ-9 of 2.91 for the population of the study (individuals ages 18 years old and above)[5]. Data collected from both studies demonstrate that depression amongst young adults have risen since the beginning of the COVID-19 pandemic.

Acquiring data on the influence of COVID-19-related lifestyle changes on various mental illnesses such as depression, anxiety, and post-traumatic stress disorder (PTSD) is crucial. Specifically, analysis of the responses gathered from surveys can be very useful. They help to identify the specific areas of mental health that are being most negatively impacted and which aftereffects of COVID-19 may be associated with them. This type of data allows social groups such as community centres and universities to enhance and cater their mental health resources to meet the needs of community members, and specifically, young adults[7]. By adapting their mental health programs, these groups can effectively respond to the psychological repercussions of COVID-19 and most importantly, help affected individuals improve their quality of mental health.

## Author Biographies

Muntaha Marjia

Muntaha Marjia is a first-year Science student at Ryerson University studying biomedical sciences. She is interested in many different areas of research within the medical field including women's health, mental health and neurodegenerative diseases.

Lucy Chen

Lucy Chen is a student at Western University and is currently doing a specialization in physiology. She is an avid clinical and biochemical research student interested in the complex integrations of biotechnology and clinical applications.

Tian Jian Gao

Tian Jian (Jenny) Gao is a second year Medical Science student at Western University, interested in studying pathology, virology and healthcare. She has co-authored a book on vaccines and the COVID-19 disease.

Dr. Austin Mardon

Dr. Austin Mardon is the co-founder of the Antarctic Institute of Canada (AIC). Dr. Mardon has earned a PhD in geography from Greenwich University. Through AIC, Dr. Mardon supports scholarly writing and academic research done by undergraduate students all across Canada.

## References

Cucinotta, Domenico, and Maurizio Vanelli. "WHO Declares COVID-19 a Pandemic." *Acta bio-medica : Atenei Parmensis* vol. 91,1 157-160. 19 Mar. 2020, doi:10.23750/abm.v91i1.9397

Silva Junior FJGD, Sales JCES, Monteiro CFDS, *et al.* "Impact of COVID-19 pandemic on mental health of young people and adults: a systematic review protocol of observational studies." *BMJ Open,* vol. 10, no. 7, 2020, pp. e039426-e039426.

Liu, Cindy H et al. "Factors associated with depression, anxiety, and PTSD symptomatology during the COVID-19 pandemic: Clinical implications for U.S. young adult mental health." *Psychiatry research* vol. 290 (2020): 113172. doi:10.1016/j.psychres.2020.113172

Kroenke, K et al. "The PHQ-9: validity of a brief depression severity measure." *Journal of general internal medicine* vol. 16,9 (2001): 606-13. doi:10.1046/j.1525-1497.2001.016009606.x

Jia, Ru, et al. "Mental Health in the UK during the COVID-19 Pandemic: Cross-Sectional Analyses from a Community Cohort Study." *BMJ Open*, vol. 10, no. 9, 2020, pp. e040620-e040620.

Jia, Ru, et al. "Young People, Mental Health and COVID-19 Infection: The Canaries we Put in the Coal Mine." *Public Health (London)*, vol. 189, 2020, pp. 158-161.

"Using Data Science to Help Tackle Mental Health Issues." 16 Mar. 2020. Web. 30 Dec. 2020.

# The best online teaching and learning practices to implement while at home

Alyssa Wu, Peter Johnson, John Johnson, Jasrita Singh, Zach Schauer, Austin Mardon

## Abstract

Billions of people around the world are facing the challenges brought on by the COVID-19 pandemic. This all started when a new coronavirus strain (SARS-CoV-2 virus) was found in patients with pneumonia-like symptoms, located in Wuhan, China (Andersen et al.,

2020). During this time of quarantine and self-isolation at home, many students are continuing to learn and finish their studies online. Students and teachers from all around the world have had to accommodate their learning and teaching styles to be fit for learning at home.

Education now heavily relies on the internet, for sending and receiving information. Online communication platforms are now being heavily used to facilitate student-teacher interactions. Teachers are also finding better ways to share and communicate course materials effectively over the internet, in order for their students to benefit and supplement their learning progress. Teachers and educators around the world have been adapting their in-person learning environments to a similar online environment as smoothly as possible, given the current circumstances of the global COVID-19 pandemic. This literature review will focus on the challenges in various learning sectors, ranging from elementary, secondary, and post-secondary education. There will also be a focus on improving teaching practices to benefit students and help them learn and engage in more material while at home.

## The Current State of Online Learning

Online platforms are now being used to recreate the human-human interactions that most students are used to having in their day-to-day learning before the lockdown measures were enforced by the COVID-19 pandemic. As internet usage increases, supplementing online learning with in-person learning, known as blended learning, is becoming more prevalent (Singh, 2003). Given the current state of the world, instructors are now having to alter their teaching styles and use more online learning.

Currently, many forms of online learning require students to be motivated,self-directed, and goal-oriented. Students will use more materials (such as readings, videos, etc.), rather than more personal interactions (such as discussions, presentations, etc.).

Communication is vital in online learning (Rapanta et al., 2020). Teachers should try to keep students on track with a schedule laying out all the materials that students should be going through weekly. This also serves to keep students motivated and engaged with the material.

However, with the decrease in direct communication with the instructor, many students find it difficult to absorb the material that they are learning. Many students are used to engaging in teacher-centred learning, where what they learn is directly influenced by the resources that the teacher provides (Emaliana, 2017).

## The Use of Online Communication Platforms

There are a wide range of online audio-video communication platforms that are now being used (eg. Zoom, Skype, Google

Classroom, Microsoft Teams, Cisco Webex etc.) to facilitate and simulate the transition of being in a classroom to being at home, with a computer in hand (Sahu, 2020). Many online communication platforms have designed new features to simplify the online teaching and learning experience. Educators can take advantage of screen-sharing and annotation functionalities, where they can prepare a presentation and some accompanying videos as a visual to help students learn while they are teaching and explaining a new idea or concept. Instead of a chalkboard/whiteboard, teachers now have the option to use digital whiteboards to write and solve problems with their students' lives. Some platforms (such as Zoom) enable students to annotate the screen as well, and it might be engaging for everyone in the class if the instructor would like to have some students volunteer and share their ideas.

Of course, a downside to utilizing some of these great functionalities is that they are limited depending on the devices that you currently own.

## Elementary and Secondary School Education

Different school boards have been designing and using their own platforms to enable easy-access for parents and students to access learning resources to support their learning (Mulenga & Marbán, 2020). Secondary schools and high schools have tried various methods to help transition their students to this new learning environment while supporting them and preparing them for their future endeavours - whether their goal is to pursue higher education, join the workforce, etc. (Wang, 2013). This period of change has not been entirely smooth along the way, and some students have reported feeling 'unprepared' in pursuing their post-secondary studies.

To support the transition from in-person school to online school, there has been an increase in making online digital libraries, containing books, worksheets and additional learning materials for parents to download for their children. However, the effectiveness of these online libraries depends on an increase of awareness to better integrate these online resources into student learning (Sharifabadi, 2006). However, despite the continuous efforts in making this transition as smooth as possible, there are definitely some limitations in being able to keep young children (especially at the elementary level) engaged in completing their school work entirely online. Many children do not have access to their own personal devices, and so, they will need to rely on parental help and the use of their devices. But as more and more parents return to work, this is not always a feasible option. Younger children need the guidance and structure provided by parents and teachers, in order to succeed. With increasing class sizes directly relating to limiting time constraints, it is very difficult for teachers to provide one-on-one support for children who need it most (Mulholland & O'Connor, 2016).

## Education in Universities/Colleges

Many professors have been hard at work, transitioning their lectures to be in a user-friendly online format for their students. With this factor in mind, there has definitely been a wide range of challenges as it is difficult to fully simulate a lecture-style learning environment through a digital screen.

Numerous institutions are starting to default to pre-recorded lectures. This is often easier for instructors in many circumstances, as it eliminates the potential glitches in technology that can occur during a live lecture with hundreds of students tuning in live.

However, from a student's perspective, there are definitely some benefits and downsides to the use of asynchronous, pre-recorded lectures. Pre-recorded lectures allow students more flexibility in watching these lectures on their own time. These types of lecture styles require a great deal of time management on the student's behalf. Taking an opposing standpoint, students tend to fall behind in these lectures, as they are often unscheduled, and up to the student's discretion on when to watch them. The use of small assessments (such as quizzes and assignments) could be added to the end of each week, as an additional motivation to keep students on track and engaged in their course work (Kamal et al., 2020).

Synchronous (or live) lectures can sometimes be more engaging for professors and students alike, but it also adds another layer of stress. Technological challenges (such as internet stability and bandwidth) can cause delays and lags, which disrupt the flow of information being transmitted from professors to students (Gillett-Swan, 2017).

## General Teaching Recommendations and Suggestions

Designing a successful online learning environment should focus on a student-centered design. Instructors should keep in mind that the content that they are putting up for students to view should be of good quality, and they should benefit the students in learning the desired material (Rapanta et al., 2020). There are many options that are readily available to help instructors develop good quality learning resources, such as podcasts, pre-recorded lectures, educational videos, article/textbook readings, etc. In addition, instructors should keep in mind the practicality of the resources that they are sharing with the class. Time is valuable for everyone, and many students have a wide range of other commitments that they have to focus their attention on -- including schoolwork. Instructors should try to filter through their resources, and assign only the necessary sections that contain the most valuable information relevant to the course material. For example, when looking at a long article, the instructor could skim through and guide their students towards the most important sections - rather than having them read the entire article. This is a more efficient method, as it saves the student some time, and also prevents distractions from having the student read a lot of extraneous material (Förster et al., 2018).

When teachers are facing these problems with their students, they can consider switching up their teaching styles to better accommodate the student's immediate struggles. Be careful not to be too pushy or persistent in achieving the goals of a given task. Sometimes, it might be better to completely switch over to another piece of music, or try a different activity. Constantly changing up the structure of the lessons can keep the students engaged for a longer period of time.

Teachers should consider creating interactive presentations and puzzles for the students to solve in class. One of the most important ideas to keep in mind while running an online class is the level of student engagement. Some teachers have opted in and out of having students on turning on their cameras, but with that, this presents another set of challenges to overcome. At home, it is harder to control the number of distractions, as every household is unique. When compared to an in-person classroom, general distractors can be easily eliminated by the teacher.

For teachers running classes with older students (grades 4+), dedicating some time to open discussion may help improve student engagement. Teachers can call upon a student who has their 'hand raised' to share their thoughts and opinions on a given question.

Another option is to utilize the "chat" feature that is embedded in many digital platforms. The teacher could recruit an assistant, who will be able to monitor the incoming chat messages during class. The assistant could serve to filter and report back on frequently asked questions and summarize general comments and remarks made by the students in the class.

## Works Cited

Andersen, Kristian G. et al. "The proximal origin of SARS-CoV-2". *Nature Medicine,* vol. 26, Mar.

2020, pp. 450-452. https://doi.org/10.1038/s41591-020-0820-9.

Emaliana, Ive. "Teacher-centered or Student-centered Learning Approach to Promote Learning?". *Jurnal Sosial Humaniora,* vol. 10, no. 2, Nov. 2017, pp. 59-70. http://dx.doi.org/10.12962/j24433527.v10i2.2161.

Förster, Natalie, et al. "Short- and long-term effects of assessment-based differentiated reading instruction in general education on reading fluency and reading comprehension". *Learning and Instruction*, vol. 56, Aug. 2018, pp. 98-109. https://doi.org/10.1016/j.learninstruc.2018.04.009.

Gillett-Swan, Jenna. "The challenges of online learning: supporting and engaging the isolated learner". *Journal of Learning Design*, vol. 10, no. 1, 2017, pp. 20-30. https://eprints.qut.edu.au/102750/.

Kamal, Ahmad A. et al. "Transitioning to Online Learning during COVID-19 Pandemic: Case Study of a Pre-University Centre in Malaysia". *International Journal of Advanced Computer Science and Applications*, vol. 11, no. 6, 2020, pp. 217-223. https://philpapers.org/rec/KAMTTO-8.

Mulenga, Eddie M. and Marbán, José M. "Prospective Teachers' Online Learning Mathematics Activities in The Age of COVID-19: A Cluster Analysis Approach". *EURASIA Journal of Mathematics, Science and Technology Education*, vol. 16, no. 2, May 2020, pp. 1-9. https://doi.org/10.29333/ejmste/8345.

Mulholland, Monica and O'Connor, Una. "Collaborative classroom practice for inclusion: perspectives of classroom teachers and learning support/resource teachers". *International Journal of Inclusive Education,* vol. 20, no. 10, Feb. 2016, pp. 1070-1083. https://doi.org/10.1080/13603116.2016.1145266.

Rapanta, Chrysi, et al. "Online University Teaching During and After the Covid-19 Crisis: Refocusing Teacher Presence and Learning Activity". *Postdigital Science and Education,* July 2020. https://doi.org/10.1007/s42438-020-00155-y.

Sahu, Pradeep. "Closure of Universities Due to Coronavirus Disease 2019 (COVID-19): Impact on Education and Mental Health of Students and Academic Staff". *Cureus,* vol. 12, no. 4, Apr.

2020. https://doi.org/10.7759/cureus.7541.

Sharifabadi, Saeed R. "How digital libraries can support e-learning". *The Electronic Library,* vol.

24, no. 3, May 2006, pp. 389-401. https://doi.org/10.1108/02640470610671231.

Singh, Harvey. "Building Effective Blended Learning Programs". *Educational Technology,* vol. 43, no. 6, Dec. 2003, pp. 51-54. https://www.ammanu.edu.jo/EN/Content/HEC/6.pdf.

Wang, Xueli. "Why Students Choose STEM Majors: Motivation, High School Learning, and Postsecondary Context of Support". *American Educational Research Journal,* vol. 50, no. 5, Oct. 2013, pp. 1081-1121. https://doi.org/10.3102%2F0002831213488622.

## Biographies

**Alyssa Wu** is an Article Writer for the Antarctic Institute of Canada.

**Peter Anto Johnson, MSc, BSc(Hons).** is a writer, medical scientist, and mental health advocate with a strong passion for healthcare. He currently holds an MSc in medical sciences and a BSc (Hons) degree from the University of Alberta and was recognized for his notable community and research achievements, being named one of Alberta's Top 30 Under 30 in 2019.

**John Christy Johnson, BSc(Hons).** is a writer, biomedical engineer, and mental health advocate with a strong passion for accessibility and community service. He currently BSc (Hons) degree from the University of Alberta and has been recognized for his notable community and research achievements, having been named one of Alberta's Top 30 Under 30 in 2019.

**Jasrita Singh** is an undergraduate student with a background in Biochemistry, Biomedical Discovery and Commercialization at McMaster University.

**Zach Schauer** is an Article Writer for the Antarctic Institute of Canada.

**Austin Albert Mardon, CM Ph.D.** is an author, community leader, and advocate for mental health. He is an assistant adjunct professor at the John Dossetor Health Ethics Centre at the University of Alberta. In the mid 80's, he founded and today still directs the Antarctic Institute of Canada, a non-profit entity based in Edmonton, Alberta. He is also an Order of Canada member and Fellow of the Royal Society of Canada.

# The Impact on the Music and Arts Community during the COVID-19 Pandemic

Alyssa Wu, Jasrita Singh, Austin A. Mardon

## Abstract

The COVID-19 pandemic was brought into the world by a novel SARS-CoV-2 virus from Wuhan, China. The music and arts industry has been heavily impacted as a result of the lockdown measures put in place by COVID-19. Choral communities, bands, orchestras, and ensemble rehearsals have all been put on hold as a precaution to limit the spread.

Nevertheless, artists and musicians have still been creative enough to showcase their passions and share them with the world. As a result of the pandemic, ensembles and groups have been still hard-at-work recording music, putting together 'virtual' performances and sharing their creations online. Virtual choir and orchestra performances have been trending all over the internet, reaching large audiences and creating high engagement with their inspirational video performances.

## Introduction

There are multiple facets in creating a complete music and arts curriculum. Children and adults all around the world participate in the arts for a variety of reasons. For many countries, music and the arts are embedded into childhood education. Some parents choose to enroll their child(ren) into specific artistic programs to gain extracurricular experience (Foster & Jenkins, 2017). The current state of the COVID-19 pandemic has put a hold on many artistic activities conducted worldwide.

59

Teachers and educators have had to adapt their teaching and student learning practices to align with the restrictions of preventing further spread of the SARS-CoV-2 virus.

## Singing Safely During the Pandemic

A major concern with the increasing transmission of COVID-19 includes choral singing and congregational singing in small, restricted spaces. The vocal community has been targeted as a cause of "super-spreading" the virus through aerosols and respiratory droplets within their environments, through mass media (Naunheim et al., 2020). However, these claims have not been credibly sourced by concrete evidence (Sataloff et al., 2020).

The challenge with tracing the spread of the virus is that it can be spread even when the patient is asymptomatic (they show no signs of infection). Many critiques have been made against singing in public gatherings, including the generation of respiratory droplets and particles that exist in exhaled human breath. Phonation, sneezing and coughing produces a higher amount of these respiratory aerosols (Naunheim et al., 2020).

It is important to note that the information that we have on the spread and transmission of the SARS-CoV-2 virus (COVID-19) is still very preliminary and heavily based on background knowledge from similar coronaviruses that were present in the past (Wei & Li, 2016). Artistic organizations have been closely adhering to public health guidelines to keep the spread of infection under control. Common practices used globally including wearing face masks (medical or non-medical), using proper Personal Protective Equipment (PPE), and physical/social distancing (Chu et al., 2020).

# Online Rehearsals

Rehearsing online together has a completely different experience than in-person rehearsing. There are limitations in being able to recreate the same experiences that are had with musicians and artists gathering in person. For example, it could be challenging to be "in sync" with your peers on an online platform. While everyone is significantly increasing their internet usage, there will always be some form of delay and lagging experienced during the calls. This makes it difficult to play or sing 'in time' with other members of the group.

Although not well studied or tested currently, there are some plausible solutions that can help bring people together during these difficult times. Protective measures, imposed by the lockdown restrictions, such as maintaining a safe distance from others, wearing masks and using personal protective equipment (PPE) can be used to help limit the spread. In addition, larger rehearsal and performance spaces can be used so that it is feasible to maintain safe distancing. Increasing the ventilation around the room, such as HEPA filtration, may potentially be effective. Air purifiers have been shown to reduce the amount of viral particles found in the atmosphere. HEPA filters are able to capture very small particles, ranging from 5 microns ($5 \times 10^{-6}$ m) or larger (Elias & Bar-Yam, 2020). Studies have shown that deep cleaning of frequently touched surfaces, including using disinfectant cleansers and ultraviolet light (UV-C) can help to eliminate bacterial and viral contamination across frequently-touched surfaces and aerosol particles in the environment. Unfortunately, as with many other disinfectants on the market, there are limitations to using solely UV-C light to clean contaminated areas. UV-C light does not effectively reach areas that are shadowed or dark. Therefore, it has been suggested that UV-C light be used in conjunction with other cleaning solutions for in-person rehearsal spaces (Dexter, 2020).

# Live-streaming Concerts and Recitals

Musicians and artists dedicate a substantial amount of time practicing their individual parts and rehearsing as an ensemble. Most often, a common goal is to work towards an upcoming concert, recital, or showcase. However, during the age of COVID-19 pandemic, many concert halls and performance venues were forced to be closed down to prevent the spread of the virus as an attempt to flatten the curve.

Online, digital pre-recorded ensemble or live-streamed solo concerts are a great way to showcase the progress that a group had been making over the season. Many planned organizational concerts were cancelled abruptly, and yet, members have been working hard on perfecting their repertoire to share with their communities. Hosting a digitalized concert can be a great way to share the progress that has been made as a group, and with enough advertising, this has the potential to reach an even larger audience that may not have been able to attend in-person concert venues.

Collaborative choirs and orchestras have also become increasingly popular during the time of the COVID-19 pandemic. Musicians have been collectively working with other musicians geographically located in different parts of the world to create music through the power of the Internet (Rofe et al., 2017).

## Teaching Lessons Online

Teaching music is also very difficult, with all the limitations set in place. Music teachers need to find additional creative ideas to keep their students engaged and build positive progress in developing their skills. As audio/video sharing is now widely used and easily accessible for most people, there has been an increase in teachers creating 'tutorial' videos or 'listening tracks' to help students maintain their effective practice at home.

There are a lot of setbacks during video lessons that can leave students feeling stressed and discouraged during a lesson, and during their own solo practice sessions at home. For example, a student learning piano may find it difficult to read the musical notation on the score. Despite their dedication and persistence, it can be hard to correct and improve these skills online.

It is also difficult to correct posture, for example, if a student is learning to play the violin. Here, proper body posture, correct positioning of the violin, and bow hold is crucial to solidify from the start. Often times, teachers find it very difficult to describe exactly how they are doing something. When lessons were in-person, many teachers would tend to gravitate towards physically altering the student's position — but this can't be easily done digitally.

COVID-19 has truly emphasized issues with inequality of access to educational resources for students and their families. At the start of the global lockdown, many organizations were offering resources either free of charge, or with a reduced rate for a limited period of time. Despite these attempts to make more resources readily accessible for everyone, some families still continue to struggle (Daubney & Fautley, 2020).

## Works Cited

Chu, Derek K. et al. "Physical distancing, face masks, and eye protection to prevent

person-to-person transmission of SARS-CoV-2 and COVID-19: a systematic review and meta-analysis". *The Lancet*, vol. 395, no. 10242, June 2020, pp. 1973-1987. https://doi.org/10.1016/S0140-6736(20)31142-9.

Daubney, Alison and Fautley, Martin. "Editorial Research: Music education in a time of pandemic". *British Journal of Music Education*, vol. 37, no. 2, July 2020, pp. 107-114. https://doi.org/10.1017/S0265051720000133.

Dexter, Franklin, et al. "Perioperative COVID-19 Defense: An Evidence-Based Approach for Optimization of Infection Control and Operating Room Management". *Anesthesia and analgesia*, vol. 131, no. 1, Apr. 2020, pp. 37-42. https://doi.org/10.1213%2FANE.0000000000004829.

Elias, Blake and Bar-Yam, Yaneer. "Could Air Filtration Reduce COVID-19 Severity and Spread?".

*New England Complex Systems Institute*, Mar. 2020, pp. 1-4.

Foster, E. Michael and Jenkins, Jade V. M. "Does Participation in Music and Performing Arts Influence Child Development?". *American Educational Research Journal*, vol. 54, no. 3, June 2017, pp. 399-443. https://doi.org/10.3102%2F0002831217701830.

Naunheim, Matthew R. et al. "Safer Singing During the SARS-CoV-2 Pandemic: What We Know and What We Don't". *Journal of Voice,* July 2020. https://doi.org/10.1016/j.jvoice.2020.06.028

Rofe, Michael, et al. "Online Orchestra: Connecting remote communities through music".

*Journal of Music, Technology & Education*, vol. 10, no. 2-3, December 2017, pp. 147-165. https://doi.org/10.1386/jmte.10.2-3.147_1.

Sataloff, Robert T. et al. "Singing and the Pandemic: Return to Performance?". *Journal of Voice*, Aug. 2020. https://doi.org/10.1016/j.jvoice.2020.07.031.

Wei, Jianjian and Li, Yuguo. "Airborne spread of infectious agents in the indoor environment".

*American Journal of Infection Control*, vol. 44, no. 9, pp. 102-108. https://doi.org/10.1016/j.ajic.2016.06.003.

**Biographies**

**Alyssa Wu** is an Article Writer for the Antarctic Institute of Canada.

**Jasrita Singh** is an undergraduate student with a background in Biochemistry, Biomedical Discovery and Commercialization at McMaster University.

**Austin Albert Mardon, CM Ph.D.** is an author, community leader, and advocate for mental

health. He is an assistant adjunct professor at the John Dossetor Health Ethics Centre at the University of Alberta. In the mid 80's, he founded and today still directs the Antarctic Institute of Canada, a non-profit entity based in Edmonton, Alberta.

# To what extent can we stave off Alzheimer's Disease? A brief review of Different treatment approaches for Alzheimer's Disease

Authors: Muntaha Marjia[1, 2], Lucy Chen[1, 3], Tian Jian Gao[1, 4], and Dr. Austin Mardon[1, 5]

**Author Affiliations:**

[1]Antarctic Institute of Canada, #103 11919 82 St NW, Edmonton AB, T5B 2W4, Canada

[2]Biomedical Sciences, Faculty of Science, Ryerson University Toronto, Ontario

[3]Honours Specialization in Physiology, Medical Sciences, Faculty of Science and Schulich School of Medicine and Dentistry, Western University, London Ontario

[4]Medical Sciences, Faculty of Science and Schulich School of Medicine and Dentistry, Western University, London, Ontario

[5]Department of Psychiatry & John Dossetor Health Ethics Centre, University of Alberta, 2J2.00 WC Mackenzie Health Sciences Centre, 8440 112 St NW, Edmonton AB, T6G 2R7, Canada.

As humans, we have a preconditioned need for physical evidence of the events and people in our lives, and we retain them as our memories. From filming embarrassing videos of friends all the way to taking excessive pictures during family parties, we document everything and sometimes without even realizing it. However, when one develops Alzheimer's disease, a simple memory task, such as knowing where the car keys are, becomes a daily challenge. Alzheimer's disease is a cause of dementia that grows with age due to the degeneration of neuron cells, which significantly impacts memory, thought processes, and behaviour[1,2]. So in the end, perhaps human nature does make sense. We are inclined to document everything because subconsciously, we do not trust our minds to protect our memories.

But what if Alzheimer's disease could be staved off? Neurologists and neuroscientists have been asking themselves this very question for decades. However, different results from recent studies have led researchers to form different conclusions regarding the future of research regarding Alzheimer's disease. For instance, the National Institute on Aging, Eli Lilly and Company as well as other organizations have funded an on-going project called the Anti- Amyloid Treatment in Asymptomatic Alzheimer's (A4) study[3]. Researchers working within the A4 study believe that by targeting patients with an early accumulation of amyloid, a product of abnormal protein aggregation which causes neural cell death, they can prevent the development of Alzheimer's before symptoms erupt. A solution that involves the use of the antibody Solanezumab was proposed as a treatment for Alzheimer's disease. Solanezumab binds to soluble amyloid-beta aggregations and is predicted to have the potential to stop its clump formation in the brain[4]. However, three 18 month trials have been unable to show a correlation between Solanzezumab use and improvement in cognitive tests amongst participants with Alzheimer's disease[5]. Scientists studying Alzheimer's disease at the University of Bristol have a much different approach. Their research focuses on the prevention of calcium ion accumulation in neurons to delay the development of Alzheimer's disease[6]. One of the ways in which calcium enters the cell is through voltage-gated calcium channels (VGCCs) which is a type of ion channel on the membrane of brain cells. It was found that amyloid-beta peptide, a pathological indication of Alzheimer's disease, causes an elevated amount of calcium to enter the cell and this in reverse increases amyloid-beta peptide levels[7]. Therefore, scientists have researched the potential use of calcium channel blockers (CCBs) which prevent calcium ions from accumulating in neurons by targeting VGCCs[7]. Following a different pathway from the previous studies, neurophysicist, Gwenn Smith, and neurosurgeon, Constantine Lyketsos, have studied deep brain stimulation (DBS) on the fornix region as a neurosurgical method to reduce symptoms and within Alzheimer patients. DBS refers to placing electrodes in the areas of interest within the brain which create electrical impulses in an effort to stimulate that part of the brain[8]. Smith and Lyketsos demonstrated that patients aged 65 and over who received DBS for one year showed better cognitive function and higher cerebral glucose metabolism ( glucose is the brain's preferred source of energy)[9]. Although all three studies share a common goal towards treating Alzheimer's disease, their various studies have led to different viewpoints regarding how to approach this complex illness.

While there has been debate over exactly what features of the brain to target in order to lessen the effects of Alzheimer's disease, researchers can all agree that the cure for this illness is yet to be found. With the rapid advancement in technology and scientific methods, new questions and possibilities arise. Is there a way for neurons to undergo mitosis so they can duplicate? Can there be such a thing as memory harvesting? It is clear that strong efforts have been made towards delaying the onset of Alzheimer's disease and mitigating the symptoms of current patients. However, completely alleviating the disease or in other words, fully restoring one's healthy brain cells, continues to be researched. So for now, these questions remain simply rhetorical.

## Author Biographies

Muntaha Marjia

Muntaha Marjia is a first-year Science student at Ryerson University studying biomedical sciences. She is interested in many different areas of research within the medical field including women's health, mental health and neurodegenerative illnesses.

Lucy Chen

Lucy Chen is a student at Western University and is currently doing a specialization in physiology. She is an avid clinical and biochemical research student interested in the complex integrations of biotechnology and clinical applications.

Tian Jian Gao

Tian Jian (Jenny) Gao is a second-year Medical Science student at Western University, interested in studying pathology, virology and healthcare. She has co-authored a book on vaccines and the COVID-19 disease.

Dr. Austin Mardon

Dr. Austin Mardon is the co-founder of the Antarctic Institute of Canada (AIC). Dr. Mardon has earned a PhD in geography from Greenwich University. Through AIC, Dr. Mardon supports scholarly writing and academic research done by undergraduate students all across Canada.

# References

Herndon, Jaime. "Everything You Need to Know About Alzheimer's Disease." 25 May 2019. Web. 01 Jan. 2021.

National Institute on Aging. *Preventing Alzheimer's Disease: What do we Know?* National Institute on Aging, National Institute of Health, U.S. Department of Health and Human Services, Gaithersburg, Md., 2012.

Servick, Kelly. "Another Major Drug Candidate Targeting the Brain Plaques of Alzheimer's Disease has Failed. What's Left?" *Science (American Association for the Advancement of Science)*, 2019.

Carlson, Christopher, et al. "Amyloid-Related Imaging Abnormalities from Trials of Solanezumab for Alzheimer's Disease." *Alzheimer's & Dementia: Diagnosis, Assessment & Disease Monitoring*, vol. 2, no. 1, 2016, pp. 75-85.

Farlow, Martin R., et al. "Solanezumab in-depth Outcomes." *Alzheimer's & Dementia*, vol. 16, 2020, pp. n/a.

"Calcium Channel Blockers may be Effective in Treating Memory Loss Associated with Alzheimer's." European Union News, 2019.

Goodison, William V., Vincenza Frisardi, and Patrick G. Kehoe. "Calcium Channel Blockers and Alzheimer's Disease: Potential Relevance in Treatment Strategies of Metabolic Syndrome." *Journal of Alzheimer's Disease*, vol. 30 Suppl 2, no. s2, 2012, pp. S269-S282.

Lyketsos, Constantine, et al. "Deep Brain Stimulation Targeting the Fornix for Mild Alzheimer Dementia: Design of the ADvance Randomized Controlled Trial." *Open Access Journal of Clinical Trials*, vol. 7, 2015, pp. 63.

Blum, Karen. "Probing Deep Brain Stimulation for Alzheimer's Disease." *Probing Deep Brain Stimulation for Alzheimer's Disease.* Johns Hopkins Medicine, 21 Mar. 2017. Web. 01 Jan. 2021.

# Perceived Isolation and Health: Does isolation and feeling of loneliness pose a risk for severe SARS-CoV-2 infection?

Lina Lombo[1], Jasrita Singh[2], Peter A. Johnson[3], John C. Johnson[4], Austin A. Mardon[5]

**Affiliations:**

Bachelor of Medical Sciences, University of Western Ontario, London, Ontario, Canada

Department of Biochemistry and Biomedical Sciences, McMaster University, Hamilton, Ontario

Faculty of Medicine and Dentistry, University of Alberta

Faculty of Engineering, University of Alberta

John Dossetor Health Ethics Centre, University of Alberta, Edmonton, Alberta, Canada

Correspondence Email: aamardon@yahoo.ca

## Abstract

The SARS-CoV-2 pandemic has led to worldwide stay-at-home orders and social isolation. Despite the lack of research exploring the potential influence of loneliness and isolation on the severity of SARS-CoV-2, studies have related loneliness and isolation to factors directly attributing to aggravating the severity of SARS-CoV-2. Although there are many factors that contribute to severe illness from SARS-CoV-2, taking measures to reduce the feeling of isolation may serve as a viable prevention measure. This review sought to examine both basic science and clinical literature and databases. Primary research articles, including case studies, and non-primary studies centered around human and animal studies regarding isolation and their health effects were additionally included. This review aims to provide an alternative perspective to isolation measures put in place by public health and bring to light the importance of mitigating isolation.

## Introduction

The SARS-CoV-2 pandemic has led to worldwide stay-at-home orders and social isolation. Social isolation has been defined as a lack of social interaction within groups and communities (Seyfzadeh et al., 2019), which has prevailed during quarantine and social distancing orders. Additionally, perceived isolation (i.e. loneliness), which is the subjective feeling of being isolated while not being physically alone, continues to co-occur with social isolation for many during this pandemic. Driven by the measures of social distancing and isolation to lower the spread of the virus, isolation and loneliness have affected many demographics. During the SARS-CoV-2 pandemic, seniors in nursing homes may be at greater risk for isolation. As most nursing homes have canceled group activities, confinement to their rooms and the lack of digital literacy among elders has exacerbated the feeling of isolation within nursing homes and other elders living alone (Simard & Volicer, 2020). Additionally, groups with a lack of technological connectivity can also feel increased loneliness during lockdowns (Okolo Sophie, 2019).

Despite the lack of research exploring the potential influence of loneliness and isolation on the severity of SARS-CoV-2, studies have shown that loneliness and isolation contribute to factors directly attributing to aggravating the severity of SARS-CoV-2 (Cole et al., 2007; Cruces et al., 2014; Hackett et al., 2012; Jaremka et al., 2013;

Kiecolt-Glaser, Garner, et al., 1984; Kiecolt-Glaser, Ricker, et al., 1984; Kim et al., 2016; Minotti et al., 2020; Novotney Amy, 2019; *Scientific Evidence for Conditions That Increase Risk of Severe Illness | COVID-19 | CDC*, 2020; Thurston & Kubzansky, 2009; Zheng et al., 2020) . These studies have demonstrated an association between isolation, lowered immune response, higher inflammation markers, cardiovascular disease, and risk for diabetes (Cole et al., 2007; Cruces et al., 2014; Hackett et al., 2012; Jaremka et al., 2013; Kiecolt-Glaser, Garner, et al., 1984; Kiecolt-Glaser, Ricker, et al., 1984; Kim et al., 2016; Minotti et al., 2020; Novotney Amy, 2019; *Scientific Evidence for Conditions That Increase Risk of Severe Illness | COVID-19 | CDC*, 2020; Thurston & Kubzansky, 2009; Zheng et al., 2020). Although there are many factors that contribute to severe illness from SARS-CoV-2, taking measures to reduce the feeling of isolation may serve as a viable prevention measure. Hence, the objective of this review was to evaluate current literature examining the association between loneliness and severe SARS-CoV-2 illness. This review aims to provide an alternative perspective to isolation measures put in place by public health and bring to light the importance of mitigating isolation.


This review sought to examine both basic science and clinical literature and databases including PubMed/MEDLINE, EMBASE, and Google Scholar were utilized with no time, setting, or language restrictions imposed on the search strategy. Primary research articles, including case studies, and non-primary studies centered around human and animal studies regarding isolation and their health effects were additionally included. Any study concerning meta-analysis was excluded from this review. For reference, studies that looked at the feeling of loneliness and health effects, determined if their subjects were lonely using the UCLA Loneliness Scale (Kiecolt-Glaser, Ricker, et al., 1984). While studies focusing on isolation, placed the test animals in physical isolation (Cruces et al., 2014).

# SARS-CoV-2 Mechanism Host physiological and Immune responses

The SARS-CoV-2 virus is known to primarily affect the lower respiratory system, and in some cases also affects other organ systems (Yuki et al., 2020). It primarily affects the epithelial lung cells in the alveolar space by binding to many Angiotensin-converting enzyme 2 (ACE2) receptors present (Yuki et al., 2020). These receptors make way for the virus capsule to enter via endocytosis. Normally, ACE2 receptors bind and break down angiotensin II(Sriram et al., 2020). Angiotensin II are biomolecules that increase blood pressure and inflammation and mediate organ damage (Sriram et al., 2020). Hence by breaking down angiotensin II, ACE2 receptors lower blood pressure and prevent tissue damage. Yet, this mechanism is inhibited when the spikes of SARS-CoV-2 bind to these receptors. Thus, receptors blocked by the virus hinders the ability of the ACE2 receptors to control inflammation and tissue damage promoted by angiotensin II, which is why inflammation and tissue damage is one of the major effects of the virus.

The first immune response to SARS-CoV-2 is an innate immune response from epithelial cells, alveolar macrophages, and dendritic cells (DCs) in the lung lining (Yuki et al., 2020). Next, the presentation of viral particles to T cells occurs, followed by the release of CD 4+ and CD 8+ T cells to activate B cells for the production of virus-specific antibodies and kill virally infected cells, respectively(Yuki et al., 2020). In non-severe cases, these antibodies would then go on and deactivate the viruses allowing them to be cleaved and eliminated by other immune cells.

In severe cases of SARS-CoV-2, this immunological response has been accompanied by thrombosis, pulmonary embolisms, and increased inflammation (Yuki et al., 2020), which leads to further organ impairment and damage (Konig et al., 2020). Additionally, the increased viral attack and inflammation in the lungs cause a rapid decrease of endothelial cells in the lungs contributing to hypoxemia (Hansen Mike, 2020). Endothelial cells are most vulnerable to apoptosis in patients with preexisting diabetes and high blood pressure (Hansen Mike, 2020).

The inflammation seen in severe cases has been driven by cytokine storms, increased levels of proinflammatory cytokines (Yuki et al., 2020). This mechanism in severe SARS-CoV-2 patients includes increased levels of Interleukin (IL)-6, IL-10, tumor necrosis factor (TNF)-α, macrophage inflammatory protein, monocyte chemoattractant protein (MCP1) and granulocyte- colony stimulating factor (G-CSF) (Yuki et al., 2020). Out of these, IL-6 has been known to be the most predictive inflammatory marker for severe illness (Yuki et al., 2020) . These proinflammatory cytokines lead to alveolar inflammation and impair the infect lung's ability to perform gas exchange in patients requiring mechanical ventilation (Yuki et al., 2020).

Additionally, several severe cases of SARS-CoV-2 have presented thrombosis and pulmonary embolisms due to increased levels of d-dimer and fibrinogen (Yuki et al., 2020).

## Loneliness and the innate immune system:

The innate immune system is significant in preparing the body for adaptive immunity and swiftly killing virally infected cells and preparing a more specific immune response through adaptive immunity such as antibodies. Several studies focused on the innate immunity of isolated groups, lonely groups versus non-isolated, and non-lonely groups have seen deleterious effects on the immunity of lonely and isolated groups. A study of elderly rats that were isolated for

8 weeks found a significantly lower natural killer (NK) cell activity and proliferation response of lymphocytes (Cruces et al., 2014). A similar finding of a decline in NK cell activity was reported by studies in lonely medical students and psychiatric inpatients (Kiecolt-Glaser, Garner, et al., 1984; Kiecolt-Glaser, Ricker, et al., 1984). NK cells are known to kill virally infected cells as part of the innate immune response and release antiviral cytokines such as IFN-$\gamma$ (Eissmann Philipp, n.d.). In fact, deficiencies in NK cells have been associated with increased viral susceptibility (Eissmann Philipp, n.d.). Hence, a lack of these cells prior to a viral infection such as SARS-CoV-2 would not be ideal as the immune system would need to rely on other immune mechanisms to fight the infection. However, some literature has suggested that immunocompromised individuals are not as prone to a severe course of infection (Minotti et al., 2020), thus this point still needs to be further researched. Nonetheless, several studies have found a positive association between loneliness, weaker immune system, and severe SARS-CoV-2 prognosis.

Furthermore, DNA microarray analysis has shown downregulation of regulatory genes supporting type I interferon responses and mature B lymphocyte function (Cole et al., 2007) in lonelier individuals. Type I interferons and B lymphocytes from part of the innate immune response. Analysis of NK cell function using peripheral blood samples from SARS-CoV-2 patients found an inverse correlation between disease severity and NK cell levels (Zheng et al., 2020).

## Loneliness and Inflammation

As noted earlier, increased severity of SARS-CoV-2 is associated with increased inflammatory cytokines. In this area, there have also been a series of studies in both animal models and humans that show a link in increased inflammatory cytokines, and loneliness and isolation. Compared to those feeling more socially connected, lonelier people have been reported to have increased MCP-1 (Hackett et al., 2012), up-regulation of pro-inflammatory genes, increased cytokines including IL-6, and TNF-$\alpha$ (Jaremka et al., 2013). Even so, psychological stressors have also been associated with higher counts of TNF-$\alpha$ and IL-6

(Jaremka et al., 2013). While the presented studies are not directly linked to the inflammation seen in severe cases of SARS-CoV-2, these studies show an increase in the same inflammatory cytokines related to severe illness from SARS-CoV-2. Similarly, high levels of these same cytokines have been related to other diseases. In all these studies, there were significantly lower levels of inflammatory cytokines in the non-lonely, and non-isolated groups.

## Loneliness and other risk factors for Severe SARS-CoV-2 Infection

Two of the highest-ranking risks for severe SARS-CoV-2 infection are pre-existing heart conditions such as coronary artery disease, hypertension, (*Scientific Evidence for Conditions That Increase Risk of Severe Illness | COVID-19 | CDC*, 2020) increased fibrinogen leading to pulmonary embolisms (Yuki et al., 2020). In a longitudinal study, loneliness was prospectively associated with coronary artery disease (Thurston & Kubzansky, 2009). Another study found similar findings, relating loneliness to a 30% increased risk of coronary artery disease (Novotney Amy, 2019). Researchers from the Framingham Heart Study have found increased fibrinogen levels in those that had been very few social contacts in comparison to those that were more connected (Kim et al., 2016). These higher levels of fibrinogen are the same ones that were found in severe cases of SARS-CoV-2 that led to pulmonary embolism and thrombosis (Yuki et al., 2020).

# Conclusion

Isolation and loneliness have been shown to play a major part in the innate immune system and inflammatory response to infection. Furthermore, several studies also show that loneliness can lead to several other comorbidities that place individuals at greater risk for severe prognosis. Although the public health measures such as the mandatory isolation and physical distancing are essential for slowing down the transmission of SARS-CoV-2, mitigating loneliness can be a step towards preventing severe prognosis. Taking steps to mitigate isolation and loneliness can ease the strains on the healthcare system, especially in countries with greater infection rates. This could be done by providing ways to connect with others visually through video calls or by providing seniors and low-income families with access to technological devices. With the lack of targeted therapy to treat this virus, it is critical to take all preventative measures to lower the risk of severe infections. Public health and governmental authorities should consider mitigating the feeling of isolation as a preventative measure for severe illness. Similarly, scientists should explore direct links between loneliness and severe SARS-CoV-2 infections.

### Acknowledgements

We would like to thank the Antarctic Institute and Austin A. Mardon for the support and guidance in writing this article.

## References

Cole, S. W., Hawkley, L. C., Arevalo, J. M., Sung, C. Y., Rose, R. M., & Cacioppo, J. T. (2007b). Social regulation of gene expression in human leukocytes. *Genome Biology*, *8*(9). https://doi.org/10.1186/gb-2007-8-9-r189

Cruces, J., Venero, C., Pereda-Pérez, I., & De la Fuente, M. (2014). A higher anxiety state in old rats after social isolation is associated to an impairment of the immune response. *Journal of Neuroimmunology*, *277*(1–2), 18–25. https://doi.org/10.1016/j.jneuroim.2014.09.011

Eissmann Philipp. (n.d.). *Natural Killer Cells | British Society for Immunology*. Retrieved August 27, 2020, from https://www.immunology.org/public- information/bitesizedimmunology/cells/natural-killer-cells

Hackett, R. A., Hamer, M., Endrighi, R., Brydon, L., & Steptoe, A. (2012). Loneliness and stress-related inflammatory and neuroendocrine responses in older men and women. *Psychoneuroendocrinology*, *37*(11), 1801–1809. https://doi.org/10.1016/j.psyneuen.2012.03.016

Hansen Mike. (2020, May 6). *What Doctors Are Learning From Autopsy Findings of New Coronavirus Patients | COVID-19 - YouTube*. https://www.youtube.com/watch?v=KzKvIYwqQkE

Jaremka, L. M., Fagundes, C. P., Peng, J., Bennett, J. M., Glaser, R., Malarkey, W. B., & Kiecolt-Glaser, J. K. (2013). Loneliness Promotes Inflammation During Acute Stress. *Psychological Science*, *24*(7), 1089–1097. https://doi.org/10.1177/0956797612464059

Kiecolt-Glaser, J. K., Garner, W., Speicher, C., Penn, G. M., Holliday, J., & Glaser, R. (1984).

Psychosocial modifiers of immunocompetence in medical students. *Psychosomatic Medicine, 46*(1), 7–14. https://doi.org/10.1097/00006842-198401000-00003

Kiecolt-Glaser, J. K., Ricker, D., George, J., Messick, G., Speicher, C. E., Garner, W., & Glaser,

R. (1984). Urinary cortisol levels, cellular immunocompetency, and loneliness in psychiatric inpatients. *Psychosomatic Medicine, 46*(1), 15–23. https://doi.org/10.1097/00006842-198401000-00004

Kim, D. A., Benjamin, E. J., Fowler, J. H., & Christakis, N. A. (2016). Social connectedness is associated with fibrinogen level in a human social network. *Proceedings of the Royal Society B: Biological Sciences, 283*(1837), 20160958. https://doi.org/10.1098/rspb.2016.0958

Konig, M. F., Powell, M., Staedtke, V., Bai, R. Y., Thomas, D. L., Fischer, N., Huq, S., Khalafallah, A. M., Koenecke, A., Xiong, R., Mensh, B., Papadopoulos, N., Kinzler, K. W., Vogelstein, B., Vogelstein, J. T., Athey, S., Zhou, S., & Bettegowda, C. (2020). Preventing cytokine storm syndrome in COVID-19 using α-1 adrenergic receptor antagonists. In *Journal of Clinical Investigation* (Vol. 130, Issue 7, pp. 3345–3347). American Society for Clinical Investigation. https://doi.org/10.1172/JCI139642

Merrett, J., Barnetson, R. S., Burr, M. L., & Merrett, T. G. (1984). Total and specific IgG4 antibody levels in atopic eczema. *Clinical and Experimental Immunology, 56*(3), 645–652. https://doi.org/10.1111/(ISSN)1365-2249

Minotti, C., Tirelli, F., Barbieri, E., Giaquinto, C., & Donà, D. (2020a). How is immunosuppressive status affecting children and adults in SARS-CoV-2 infection? A systematic review. In *Journal of Infection* (Vol. 81, Issue 1, pp. e61–e66). W.B. Saunders Ltd. https://doi.org/10.1016/j.jinf.2020.04.026

Novotney Amy. (2019, May). *The risks of social isolation.*

https://www.apa.org/monitor/2019/05/ce-corner-isolation

Okolo Sophie. (2019, June 26). *Loneliness is an epidemic, and we can turn to technology to fix it.* https://massivesci.com/articles/loneliness-technology-older-adults-aging-internet- accessapps-mind-control/

*Scientific Evidence for Conditions that Increase Risk of Severe Illness | COVID-19 | CDC.* (2020, July 28). https://www.cdc.gov/coronavirus/2019-ncov/need- extraprecautions/evidence-table.html

Seyfzadeh, A., Haghighatian, M., & Mohajerani, A. (2019). Social isolation in the elderly: The neglected issue. In *Iranian Journal of Public Health* (Vol. 48, Issue 2, pp. 365–366). Iranian Journal of Public Health. https://doi.org/10.18502/ijph.v48i2.844

Simard, J., & Volicer, L. (2020). Loneliness and Isolation in Long-term Care and the COVID-19 Pandemic. In *Journal of the American Medical Directors Association* (Vol. 21, Issue 7, pp. 966–967). Elsevier Inc. https://doi.org/10.1016/j.jamda.2020.05.006

Sriram Krishna, I. P. R. L. (2020, May 14). *What is the ACE2 receptor, how is it connected to coronavirus and why might it be key to treating COVID-19? The experts explain.* https://theconversation.com/what-is-the-ace2-receptor-how-is-it-connected-to- coronavirusand-why-might-it-be-key-to-treating-covid-19-the-experts-explain-136928

Thurston, R. C., & Kubzansky, L. D. (2009). Women, loneliness, and incident coronary heart disease. *Psychosomatic Medicine*, *71*(8), 836–842. https://doi.org/10.1097/PSY.0b013e3181b40efc

Yuki, K., Fujiogi, M., & Koutsogiannaki, S. (2020). COVID-19 pathophysiology: A review. In *Clinical Immunology* (Vol. 215, p. 108427). Academic Press Inc. https://doi.org/10.1016/j.clim.2020.108427

Zheng, M., Gao, Y., Wang, G., Song, G., Liu, S., Sun, D., Xu, Y., & Tian, Z. (2020). Functional exhaustion of antiviral lymphocytes in COVID-19 patients. In *Cellular and Molecular Immunology* (Vol. 17, Issue 5, pp. 533–535). Springer Nature. https://doi.org/10.1038/s41423-020-0402-2

# The Canadian Public Health Perspectives on Quarantined Primary Education of 2020- 2021: A COVID-19 Retrospective Review

**Authors**: D. Bhavsar[1], J. Singh[1], A. Mardon[2]

Affiliations:

Department of Biochemistry and Biomedical Sciences, McMaster University, Hamilton, Ontario, Canada

John Dossetor Health Ethics Centre, University of Alberta, Edmonton, Alberta, Canada

**\*Correspondence Author:**

Daivat Bhavsar

Department of Biochemistry and Biomedical Sciences, McMaster University, Hamilton, Ontario, Canada

Tel: 647-680-7086

E-mail: bhavsd5@mcmaster.ca

**Running head**: Canadian schools re-open during COVID-19

## ABSTRACT:

As September approached and schools prepared to re-open, the mental health and developmental needs of students and staff had become a rising concern. The bettering epidemiological status of the coronavirus pandemic in the Summer of 2020 had encouraged all stakeholders of education to push for in-person classes. Provincial authorities had published accessible plans to facilitate social bonding and development through direct interactions in a safe environment, something that renders distance education as an inadequate substitute for learning and development. However, a large parent body also served an opposing force to question the level of safety associated with the 'enhanced' protocols implemented that fall, which often led authorities to reconsider specific decisions regarding ventilation, cleaning measures, and the organization of student groups in schools. This review paper aims to summarize the key progressions of the education ministers of British Columbia, Alberta, Quebec and Ontario by analysing the official plans and statements for this year's primary education. Other public health perspectives from parents and staff (unions) were also investigated through polls, petitions, and protests. Finally, a rapid review study by McMaster University was considered to suggest that students may have been in safe hands based on the success of the reopening of schools in other nations and that schools are not to be entirely blamed for the second wave.

## INTRODUCTION:

Six months. The provincial government and public health authorities of Canada had almost six months to carefully organize primary education during the coronavirus (COVID-19) pandemic for the 2020-2021 school year. Schools and universities had shifted to distance education as early as March when the World Health Organization (WHO) had declared the coronavirus as a pandemic on March 11th [1]. Provinces began declaring a state of emergency due to the increasing spread of COVID-19 in Canada. Since early July, the COVID-19 outbreak curve had greatly 'flattened', as confirmed by Canada's top doctor, Dr. Theresa Tam [2].

Consequently, most provinces announced that schools were aiming to safely hold in-person classes. The key rationale for this decision was to promote the physical and mental health of students as direct peer interactions and hands-on learning in classroom settings are believed to be crucial for development [3], especially for students with identified special needs. Yet, these decisions met with much public uneasiness. Parents wanted to know 'how safe is the September' for their children that were expected to return to school [4].

This retrospective review exclusively focuses on the four most-affected provinces: Ontario, British Columbia (BC), Alberta, and Quebec. Education plans at the primary level will be analysed for these selected provinces from a public health perspective. The analysis was conducted based on secondary research findings from mostly news articles published under renowned agencies, governmental reports and guides, and professional reports/studies by educational institutions. The paper seeks to investigate the most safe and effective ways of balancing the challenges of social interaction whilst preventing transmission and school outbreaks.

# PRIMARY EDUCATION – REOPENING SCHOOLS:

Satisfying multiple stakeholders of education – parents, staff, students, district boards – proved to be a major challenge for all provincial authorities across Canada. There were increasing pressures [5] for remote school options as a large proportion of parents show low comfortability with their children returning to school in September. On August 18th 2020, a new poll conducted by Leger and the Association of Canadian Studies captured the societal perspective on primary education for the coming year [6]. Results showed that 66% of parents were "worried" about schools reopening and 69% asserted that in-person classes should be "suspended"; only 19% are for in-person classes [6, 7]. However, this was an online poll that may be subject to bias as the sample of respondents is not random.

Despite the heightened anxiety about school reopening decisions, all stakeholders expressed concern for student mental health, and social engagement in the learning process [8]. Although distance education was obviously the safer option, in-person classes promote active learning through direct interactions with teachers and classmates [9]. Several studies have suggested social interaction to be fundamental to the child development process, as they continue to build on their communication and interpersonal skills [3]. Without schools, online learning requires serious commitment from parents and students to gain knowledge through a computer, and nurture peer interactions; this may not be an option for all families. Based on these concerns of mental wellbeing, provincial authorities stressed the importance of preserving the quality of professional education by planning, and publishing guides, for safe returns to in-person classes.

## British Columbia:

The provincial authorities of British Columbia launched an accessible 'Back to School Plan' webpage [10] that summarized the "new health and safety measures" put in place for in-person learning last September. Masks were only mandatory for grades 4-12 students according to the plan, which has raised parental concerns for younger students. Rob Fleming, the education minister, announced an investment of $45.6 million for the enhanced safety measures of hygiene/handwashing stations, reusable mask supplies, and additional staff recruitments for cleaning. This budget also included $3 million for remote learning and technological loans, which suggests that the province has considered the possibility of switching to distance education if needed. Authorities emphasized the concept of learning groups as the key strategy for minimizing transmission. These cohorts served to cap the maximum number of people (i.e. 60 for elementary students) the children interact with to encourage sufficient peer interactions,

[11] whilst allowing for better distancing and contact tracing measures. However, the provincial plan had a fair share of gaps and contradictory implications. Although physical distancing strategies were listed, the plan failed to answer parental concerns of how physical distancing can be maintained in a classroom of 20-25 students, within a cohort, or in a school bus; the plan did not thoroughly discuss ventilation system improvements and contingency planning in the event of an outbreak. Consequently, parents voiced their suspicions to these protocols and call for optional classes through a petition that has gained over 42 000 signatures [12]. This led school boards to push back the initial reopening date of September 8th to allow staff and faculty to discuss healthier designs for restarting schools.

## Alberta:

Similarly, Alberta Education had published an accessible 23-page re-entry plan for the general public (last updated on July 21st) [13]. This guide showed more flexibility than other provinces such as BC, Ontario, and Quebec as it considered three different scenarios of 2020-2021 education: full in-person classes, partial in-person classes, or full distance education. Adriana LaGrange, the education minister, had announced that school authorities were expected to resume in-person classes to facilitate social interactions. The plan claimed to respect "school authority autonomies at the local level" as a certain school authority may switch to another 'scenario' given an outbreak or high demand. Again, mask-wearing was not mandatory for students in grade 3 and below and students are expected to be grouped in 'learning cohorts'.

Moreover, PSA techniques including posters and animation videos [14] were implemented in schools to educate students and promote hygiene practices. Alberta Education's provision and emphasis of using a daily self-screening questionnaire, and guidance documents for playgrounds, fitness, and libraries seemed to gain greater trust among Albertans. Protests and petitions had not been as prominently visible compared to those in BC and Ontario. Rather, an anti-mask petition that argued the unnecessary mandate for grades 4-12 student mask-wearing has gained some popularity with over 2000 signatures [15]. The above-mentioned steps may suggest that Alberta had organized its education system plans quite satisfactorily (compared to BC) from a public health perspective. Not only had Alberta Education presented flexible contingency plans for local school authorities, but also may be seen as proactive with standardized content to guide daily student affairs [14]. Yet, there was a similar increasing pressure for government authorities to address classroom sizes, ventilation system, and staffing needs as September approached [16].

## Quebec:

Quebec was the hardest hit province last Summer with the highest number of infection cases, deaths, and current active cases in Canada. However, the provincial authorities asserted that with the improving epidemiological situation, all students in elementary (grade) 1, 2, and 3 would return for full in-person classes in the fall [17]. An official back-to-school plan was unveiled on August 10th that discussed mask-wearing protocols, physical distancing measures, and recreational activities. The plan also mentioned emergency protocols in the event of a school outbreak which would lead to school closure and online pedagogical support. An infographic sheet [18] conveniently summarized mask-wearing rules: not mandatory for staff and students in the classroom, not mandatory for elementary 1-4 students anywhere, and mandatory for elementary 5-6 students for facilities and other common areas when applicable (not eating, playing a sport). Similar to the other provinces discussed in this paper, the use of stable groups (aka learning cohorts) was enforced in classrooms, but the lack of physical distancing and mask- wearing protocols raised questions on the associated safety. The webpage [19], however, did not sufficiently address the funding towards enhanced safety protocols such as hand hygiene stations, staffing needs, ventilation systems or classroom sizes. Due to these gaps and the inflexibility of the plan with 100% of students expected to return to schools, staff unions and parents organized petitions to reconsider the plans [20] after identifying fundamental contradictions with other governmental measures [21]. Some unanswered questions included the lack of social distancing within classrooms and stable groups and the large classroom sizes contradicting with the allotted 10 people-per-household or public mask-wearing rules. Early protests in June and July about the mask by-laws in Quebec pointed to the hypocrisy of schools not mandating masks.

## Ontario:

After Quebec, Ontario had been greatly harassed by COVID-19. Since the end of July, Ontario showed great improvement in epidemiological statistics with fewer than 100 cases almost every day with daily numbers once reaching 500-600 cases, with a highest of 640 cases on April 24th [22]. Shortly after British Columbia announced its plan to reopen schools, the Ontario Government followed with its decision to resume full in-person classes with similar mask- wearing policies and learning cohorts as the above three provinces – mandatory for only grades 4-12 students and staff [23]. According to a poll in June by Nanos Research and Ontario Public School Boards' Association [24], only 53% of the surveyed Ontarians stated that they were at least somewhat comfortable with their children returning to in-person classes this fall and more than ten percent would not send their children back to school. Yet, a large majority (60-80%) of respondents called for greater student support that prioritizes mental health and funding for special education. Although this poll was conducted shortly before the transition to stage 2 re- openings, Education Minister Stephen Lecce faced much backlash when back-to-school plans were announced. Despite promising funding of $309 million for an additional 1300 custodians, 500 public health workers, and cleaning supplies and PPE, twitter trends of #FireLecce and #UnsafeSeptember followed [25]. A petition on Change.org [26] that called the plan 'reckless' amassed over 230,000 signatures. Moreover, the Ontario Parent Action Network (OPAN) collectively designed a mock class [27], naming it "[Doug] Ford's COVID classroom" to clearly argue how the government failed to creatively consider distancing measures and class sizes. In the second week of August, funding was increased to $500 million to improve ventilation systems and physical distancing in classrooms. Although provincial authorities attempted to reassure parents of strict safety and health protocols, the updated plan emphasized the option of voluntary school attendance and concomitant support in online learning [28]. Unions and parents remained unconvinced and unsatisfied with this solution, arguing that physical distancing and safety cannot be ensured with large class sizes of 30 students.

## GENERAL ADDITIONS:

Safety and the prevention of a prospective second wave in Canada was a priority and the main argument against the reopening of schools. There was much concern about the possibilities of children, especially asymptomatic, transmitting the disease to one another, and subsequently, their teachers, parents, and grandparents. Anxiety levels rise in anticipation of this chain reaction as September approaches.

The governments of the four provinces discussed in this paper were nevertheless encouraging in-person classes to uphold the mental and physical wellbeing of children in a professional environment. According to the above-mentioned protests and petitions for each province, the key problem was that officials struggle to fill in the gaps with details of large class sizes, successful implementation of physical distancing, ventilation systems, mask- wearing policy contradictions, and contingency planning for distance education. These unanswered questions and government statement contradictions failed to decrease the fragility of the primary education system.

However, a rapid review [29] by McMaster University School of Nursing, funded by the Public Health Agency, suggests that young children in schools and daycares are effective, but not sources of transmission. The conclusion of this paper is based on an analysis of 33 contact tracing single and synthesis studies which consistently showed "very limited transmissions" in schools and daycares. These 33 studies are primarily from European countries such as Sweden, France, and the Netherlands. Many of these studies were conducted before preventative measures were implemented in the nations; but, this paper attempts to use a tactical approach to carefully observe and learn from the school openings has worked for other nations to decide on, and possibly guide, school reopening plans in Canada. Although some studies were reported to be of low or moderate qualities by reviewers using standard critical appraisal tools [29], the findings of low transmission in these fragile settings are consistent and potentially game-changing for Canada's primary education this year.

## CONCLUSIONS:

Provincial authorities across Canada continue to search for solutions to overcome the challenges of minimizing the risks of transmission amid a worsening second wave without compromising the quality of primary education and its developmental outcomes. There is a public demand for more consideration of safer re-opening strategies, as in-person learning is especially crucial for kids with disabilities, and complex learning or special needs. Despite the enhanced safety protocols relating to class cohorts, physical distancing, school cleanliness, and mask-wearing, parents remained skeptical of the government plans' effectiveness in controlling the pandemic, and the potential second wave that is anticipated by experts. There is, and always will be, room for improvement on the control of the COVID-19 pandemic. A major limitation of this review paper is the topic of primary education for 2020-2021 is very dynamic. Public opinion and even provincial approaches were changing, somewhat messily and contradictorily, as September approached and schools began. This review only captures a part of the planning process for 2020-2021 for only four provinces – British Columbia, Alberta, Quebec and Ontario – to judge preparedness. Common aspects and documents, such as mask-wearing policies and classroom cohorts, were analysed in attempts to minimize bias. Common sources such as governmental guides and plans on webpages, local news articles, and petitions were utilized to gauge the public health perspectives. Further studies should be recommended to analyse the epidemiological situations of more provinces and to also explore the public health perspectives more in-depth after the second wave had risen to now predict the verdicts for the 2021-2022 school year.

**Acknowledgements:**

The authors would like to sincerely thank Professor Catherine Mardon for her supervision and support during the construction of this paper.

**Financial Support:**

This paper received $300 in grant funding from the TakingItGlobal charity allocated for the publishing of this article in a professional publication source.

**Declarations of Conflict of Interest:** None

Data sharing is not applicable to this paper because no new data was created or analyzed.

Daivat Bhavsar, BSc (McMaster University) is an undergraduate student with a background in Biochemistry. Jasrita Singh, BHSc (McMaster University) is an undergraduate student with a background in Biochemistry, Biomedical Discovery and Commercialization. Austin Albert Mardon, CM, FRSC (University of Alberta) is an adjunct professor in the Faculty of Medicine and Dentistry, an Order of Canada member, and Fellow of the Royal Society of Canada.

**References:**

World Health Organization. Coronavirus Disease (COVID-19) - events as they happen: 2020, 22 August 2020.

Belmonte, L. Dr. Tam Confirms Canada Flattened The Curve As She Marks Her 3rd Anniversary As Top Doc. Narcity 2020; 22 August.

Ontario Government. Guide to reopening Ontario's schools: 2020, 14 August 2020.

Berthiaume, L. Trudeau unsure about sending his kids to school, as poll suggests he's not alone. Toronto Star 2020; 18 August.

Small, K. Dozens protest Alberta's back-to-school plan outside education minister's Red Deer office. Global News 2020; 18 August.

Johnstone, H. (2020). Can't decide whether to send your kids back to school? Here's what the experts say. CBC News 2020; 22 August.

Leger's Weekly Survey - August 18, 2020 (https://leger360. com/surveys/legers-weekly- survey-august-18-2020/). Accessed 22 August 2020.

Ontario Government. Archived - Approach to reopening schools for the 2020-2021 school year: 2020, 13 August 2020.


Mahoney, J. Remote learning for elementary students poses unique challenges. The Globe and Mail 2020; 18 August.


British Columbia Government. B.C.'s Back to School Plan: 2020, 22 August 2020.


Hager, M. Start of school for B.C. students is pushed back from Sept. 8, education minister says. The Globe and Mail 2020; 22 August.


Change.org Petition. Keep "Return to School" in BC on optional or voluntary basis in September 2020 (https://www.change.org/p/adrian-dix-minister-of-health-keep-bc-schools- reopening-on-optional-or-voluntary-basis-in-september- 2020?signed=true&fbclid=IwAR-0oXqBSI2PiyNiRfubQaGMhLXJaCXNr_TfMVhWN20K4zGx Zjm-dZOw3ETX0). Accessed 22 August 2020.


Government of Alberta. 2020-2021 School Re-Entry Plan: 2020, 14 August 2020. Alberta Education 2020; pp. 4-9.


Alberta Education. K to 12 school re-entry: 2020, 22 August 2020.

Castillo, C. Petition against mask mandate in schools concerning to Alberta teachers, health expert. Global News 2020; 22 August.

Vernon, T. UCP government faces continued pressure over Alberta school re-entry plan.

Global News 2020; 18 August.

Quebec Government. Frequently Asked Questions – Education: 2020, 19 August 2020.

Quebec Government. At school, I protect myself and others!: 2020, 16 August 2020.

Quebec Government. Back-to-school plan for the Fall of 2020 (COVID-19): 2020, 22 August 2020.

Change.org Petition. Réviser (encore) le plan scolaire du Québec / Revise (Again) Quebec's School Plan (https://www.change. org/p/gouvernement-du-qu%C3%A9bec- r%C3%A9viser-le-plan-scolaire-du-qu%C3%A9bec-revise-quebec-s-school-plan-for-covid-19?recruiter=315743015&recruited_by_id=4e4a0f30-129d-11e5-a4c1- 55d413ff0321&utm_source=share_petition&utm_medium=copylink&utm_campaign=petition_d ashboard). Accessed 18 August 2020.

Tomesco, F. Quebec's back-to-school plan leaves questions unanswered, teachers say.

Montreal Gazette 2020; 11 August.

CTV News.ca Staff. Tracking every case of COVID-19 in Canada ( https://www.ctvnews.ca/health/coronavirus/tracking-every-case-of-covid-19-in-canada- 1.4852102). Accessed 22 August 2020.


Ontario Government. Ontario's Action Plan 2020: Responding to COVID-19: 2020, March 25 2020.


Nanos Research, Ontario Public School Boards Association. (2020). Parents' comfort levels with return to school are mixed, but they support funding for mental health, online learning and technology, and health and safety: 2020, June 2020.


Vaughan J. Ford blasts Ontario teachers' unions, delivers plan to reopen schools. The Post Millennial 2020; 14 August.


Change.org Petition. Ontario Demands Better: Reduce Class Sizes to Keep Schools and Communities Safe (https://www.change.org/p/ontario-demands-better-reduce-class-sizes-to-keep-schools-and-communities-safe?redirect=false). Accessed 20 August 2020.


Macdonell, B. Mock classroom shows students 'elbow to elbow' as parents protest Ontario's school plans. CTV News 2020; 12 August.

Bowden, O. Ontario education minister 'unlocks' $500M to improve distancing, ventilation for back-to-school.
CBC News 2020; 13 August.


National Collaborating Centre for Methods and Tools. Rapid Evidence Review: What is the specific role of daycares and schools in COVID-19 transmission?: 2020, July 31 2020.

# The illogic of logic: How do we understand it?

Peter A. Johnson[1], John C. Johnson[2]

**Affiliations:**

[1]Faculty of Medicine and Dentistry, University of Alberta

[2]Faculty of Engineering, University of Alberta

**Correspondence E-mail**: paj1@ualberta.ca

The Oxford English Dictionary defines **logic** as "reasoning conducted or assessed according to strict principles of validity."[1] However, the study of logic is vast. There is no clear- cut universal agreement on what logic truly is and it is becoming used casually and interchangeably in various contexts blurring our understanding of this conception.

In order to evaluate its role in society, we must begin by posing the question what exactly constitutes logic and what these "strict principles of validity" are. Before identifying distinguishing features or attempting to operationalize this concept, it is of utmost importance to consider classical ideas and perspectives that have influenced society and history. It is also essential to consider the presence of logic in our modern world, where logic is often challenged by the "illogical".

## Classical and Non-classical Logic

In *c.* 330 BC, early logicians such as Aristotle had effectively described simple semantics and modelled propositional and predicate logic in *Organon*.[2] Aristotelian views on nature and the *logos* have permeated much into the 21st century. **Propositional logic**, also known as **zeroth- order logic** entails a purely binary qualitative system where a certain claim is true or false whereas **predicate (first-order) logic** employs quantifiers and algebraic variables. An example of zeroth-order logic would be a claim with a subject and predicate, such as "Socrates is a man". In first-order logic however, the claim should necessarily require a quantified variable to reach a conclusion. In this particular example, (1) Socrates is X, (2) X is a man, and therefore Socrates is a man. In both cases, the conclusion is that a particular claim is true or false.

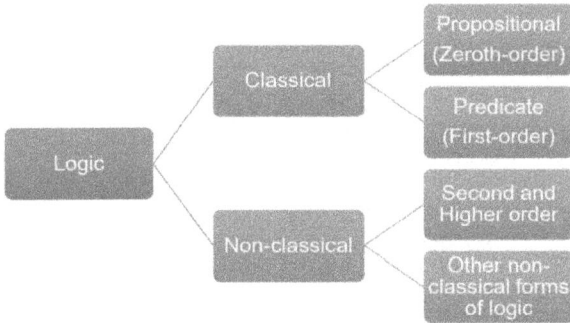

In **classical logic**, there exist "strict principles of validity", as previously defined, which are mathematical in nature. In Dov Gabbay's *Handbook of Logic in Artificial Intelligence and*

*Logic Programming*, he establishes five laws governing classical logic. All of these laws define **truth conditions**, which can then be used to deduce a conclusion. Critically however, there are major flaws in implementing classical logic in the real world.

Numerous exceptions, extensions, and variabilities in these laws arise in multiple scenarios presented in the real world and as such, various forms of **non-classical logic** that consider for these have emerged. For example, in **many-valued logic**, as opposed to two truth values – true or false – there exist other values such as unknown, infinite, etc. With the development of novel forms of non classical logic, our society has evolved from addressing theoretical problems to inferring solutions to real-world problems. Instead of scientific principles or mathematical laws, political theories and philosophical thought was emphasized.

### Non-classical Logic and Logical Pluralism

In contemporary society, **logical pluralism** is a prevailing concept that proposes there exists more than one correct logic.[4] This theory apprehends an underlying assumption that there are different standards or "principles of validity" that a logically correct conclusion must follow. Perhaps the political landscapes of the past and of our time can serve as a prototype of logical pluralism and non-classical logic in the real world.

According to the doctrines of **normative political philosophy**[4], which is concerned with how society ought to run and be, logic can be deleterious in the real world as logical proofs can provoke context-insensitive and narrow-mindedness and stands in stark opposition to the ideology of logical pluralism. The tendency of logic to reduce a problem to a single solution, especially when based on only a particular set of standards or a certain context, can be deceiving. This is patently demonstrated in the contemporary form of governance in several nations – **democracy**. Supposedly, this ideology determined that the will of the people (i.e. the majority) would result in the best decisions for society. However, are the decisions that the people, as a whole, make necessarily more informed and logical than the decisions made by one informed individual?

When considering the socio-political discipline, many recognize the significance of a "kind of logic", which has not been explicitly defined. This is consistent with the view of logical pluralism and suggests there are more forms of non-classical logic, which has yet to be explored in different contexts or under other "strict principles of validity".

## Contemporary Examples of Logic and Illogic

In the 21st century, we observe logic and illogic on multiple occasions. In fact, the entirety of society is a walking example of both of these concepts. Consider the laws governing us. Many of these laws are a product of the ideas and reasoning of informed individuals whose assertions are rooted in logic and can be visibly seen and accepted by others. Perhaps the crosstalk between the scientific community and political domain can attest to this. For instance scientific research has shed light on the disastrous calamities of climate change and global warming. Utilizing this knowledge as a premise, various legislations and emergency measures were taken to combat this communal threat. In 1992, the Kyoto Protocol[5] was based on two logical premises: (1) global warming exists and (2) human-made carbon dioxide emissions have caused it. In response to this, the steps taken internationally, nationally and communally each differed illustrating illogic. Rather than basing decisions on logic, several global citizens chose to deny the premises, act on emotion or found other rationale demonstrating logic's permanent counterpart – illogic.

Perhaps a more modern example of illogic in society are the decisions of the 2015 International Monetary Fund (IMF) and World Bank Conference at Peru five years ago[6], where numerous extractive projects are supported by the IMF. However, while the investments and business dealings were lucrative and quite profitable, there was intense conflict between displaced communities and locals. In reply, IMF and World Bank officials cited the growing global price of mineral commodities much to the logical agreement economically. However, from a quality of life standpoint, this was illogical. Not only were more people becoming poorer, the disparity in wealth also widened permitting for a few to aggregate the wealth. According to Stiglitz, a former World Bank economist and present critic, "[i]nequality is a choice — not the result of inevitable economic laws".[6] Could this be true? If it were, this would mean that the economic laws set out were not based on logic. If not, certain forms of logic must be inept in certain contexts.

One of the biggest illogical issues of the today's world may be our present election system. Since the early ages, pure democracy was unrealistic and illogical as it would require the participation of an entire population in the state, an equal representation of all people and for the vote to be based on a single standard or "strict principles of validity". During the 2016 Presidential Election, the United States of America elected Donald J. Trump as their president elect under a so-called democracy. However according to multiple sources[7,8,9], this election had a voter turnout approximately quarter the population of the state, used the first-past-the-post system voting and was based on multiple standards of logic, if at all based on logic.

All of these observations in the real world suggest that the strict principles established by logic in the laws of science and mathematics are often insufficient to address complex problems. Instead, non-classical logic and novel theories must be identified to resolve these issues.

Unfortunately, logical pluralism has made it such that there are no "strict principles of validity" in various contexts and as such, the solution is not always the same. Logic can therefore act as a double-edged sword as its reputed value for tackling real-world issues can be deceiving.

**References**:

"Logic: Definition of Logic by Lexico." (2019) Lexico Dictionaries, https://en.oxforddictionaries.com/definition/logic.

Smith, R. (2017) "Aristotle's Logic." Stanford Encyclopedia of Philosophy, Stanford University, https://plato.stanford.edu/entries/aristotle-logic/.

Gabbay DM, Hogger CJ, and Robinson JA. (1994) *Handbook of Logic in Artificial Intelligence and Logic Programming*, volume 2, chapter 2.6. Oxford University Press.

D'Agostini F. (2015) Logic and Politics. Universal Logic, State University of Milan. https://www.uni-log.org/t5-politics.html

"Kyoto Protocol to the United Nations Framework Convention on Climate Change." (1992) *United Nations Framework Convention on Climate Change*, United Nations Framework Convention on Climate Change. http://unfccc.int/resource/docs/convkp/kpeng.html

"Thousands Reject the Extractivist Logic at the World Bank-IMF Meeting in Peru." (Feb 2019) *Waging Nonviolence*, http://wagingnonviolence.org/feature/thousands-reject- extractivist-logic-world-bank-imf-meeting-peru/

Lopez. "Trump Was Elected by a Little More than a Quarter of Eligible Voters." *Vox*, 10 Nov. 2016, http://www.vox.com/policy-and-politics/2016/11/10/13587462/trump-election-2016- voter-turnout

Nye, Joseph S. "Donald Trump's Emotional Intelligence Deficit." *Al Jazeera*, Al Jazeera, 9 Sept. 2016, http://www.aljazeera.com/indepth/opinion/2016/09/donald-trump-emotional- intelligence-deficit-160907105236277.html


Samir Chopra. "A Trump Win And The First-Past-The-Post System." *Samir Chopra*, 21 July 2016, https://www.samirchopra.com/2016/07/21/a-trump-win-and-the-first-past-the-post- system/.

# Ancient Cosmology In the World's Three Monotheistic Religions.

James Fisher[1] Peter Johnson[1,2], John C. Johnson[1,3], Svetozar Zirnov[1], Daniel Polo[1] Austin Mardon[1,4].

**Affiliations:**

[1]The Antarctic Institute of Canada (11919- 82 Street NW, Edmonton, Alberta, Canada

[2]Faculty of Medicine and Dentistry, University of Alberta

[3]Faculty of Engineering, University of Alberta

[4] John Dossetor Health Ethics Centre, University of Alberta, Edmonton, Alberta, Canada

**Correspondence Email:** aamardon@yahoo.ca.

## Introduction:

In the ancient times people from various nations who have followed one of the three monotheistic religions in the world, namely Judaism, Christianity, and Islam, have held different views of cosmology than the ones commonly accepted now. Since the followers of the world's three monotheistic religions held such beliefs, they have written it into their holy books, which are used by then today for both guidance and spiritual inspiration. Being monotheistic means believing that there is only one God, and that he alone is to be worshipped, and nothing else may be worshipped beside him, since its considered idolatry. The Jewish faith holds the Tanah(The old testament of the Bible) as their source of guidance and spiritual inspiration, the Tanah is made from three parts, The Torah(The five books of the prophet Moses), the Neviim (The books of the prophets of Israel), and Ketuvim (The book of Psalms). The Tanah tells the story of the development of the Jewish nation, and the relationship it has with God. The Christian faith looks upon the new testament of the Bible as its source of guidance and spiritual inspiration. The new testament describes the life, death, resurrection, and the various miracles performed by God's son Jesus Christ. The Muslim faith aka Islam holds the words of the Quran as its source of guidance and spiritual inspiration. The book tells the story of the faith's founder the Prophet Mohammed, and the message he has been taught by the Archangel Gabriel. According to the three mono- theistic religions, the earth was flat in shape and en- closed, thus the sun and the moon were enclosed inside the earth, underneath a dome which covered the earth. Accordingly, the stars were being viewed as simply being lights in the firmament. According to their view the part of the earth where humans live was not moving, and thus motionless, but the upper part where the sun, moon, and clouds are, was moving. Also, it is important to note that the earth was viewed as being motionless, thus it was the sun and moon that orbited around the earth, and not the other way around. Even though, the three religions differ on many issues, it must be noted that their views of cosmology have been very similar, thus showing that all three religions held a different view of cosmology that the one commonly accepted nowadays.

## Research:

Since cosmology has been viewed differently in the ancient days, the views of the ancient people were preserved and made it to our days, by being recorded in the holy scriptures of the three monotheistic religions. In the Jewish faith which believes the Tanah, or old testament of the Bible is the word of God (Jehovah), there are many reference where the views of cosmology presented are different from those that are commonly accepted today. [3] In the book of Isaiah 40:22 it says "He sits enthroned above the circle of the earth, and its people are like grasshoppers. He stretches the heavens like a canopy, and spreads them out like a tent to live in". Thus, according to this passage it is clearly described that the earth was seen as a circle, and the heavens were "tent like", thus it represents the dome stretching and covering the earth's surface. When the people of Israel were in a battle with a nation called the Amorites, the prophet Joshua commands the sun to stand still in the passage of [2] Joshua 10:12 "Sun, stand still over Gibeon, and you moon, over the valley, of Aijalon". The book of Proverbs of King Solomon introduces a similar shape of the earth, as it says[3] in Proverbs 8:27"When he established the heavens, I was there; when he drew a circle on the face of the deep,", and also in [3] Proverbs 30:4 "Who has established all the ends of the earth?", those passages portray a similar image where God created the work in a shape similar to a pancake, flat and round (a circle), as well as it shows that the earth has its ends, which would be dis- proven today, since the earth is a globe, and has no ends. Similar passages are portrayed in other books of the old testament, in the book of Exodus it says[3]"The waters under the earth", thus indicating that the ancient thought that the waters from the various seas and oceans in the world were gathering up underneath the part of the earth that we live on. In the book of the prophet Daniel 4:10-11, it also says [3]"... a tree of great height in the centre of the earth...reaching with its tops to the sky and visible to the earth's furthest bounds". Thus, this passage is also describing that the earth has a centre, a middle point in the middle of it, and it describes a tree standing there and being visible to the far bounds of the earth, thus the passage ex- presses that the ancients believed that from that centre of the earth, people were able to see all the various parts of the earth, continents, seas, and oceans from one single location. In the Christian faith, the new testament of the Bible is believed to be the word of God, and it likewise presents a similar view of cosmology in its various passages. In the book of the apostle Mat- thew 4:8, it says[3]" Again, the devil taketh him up unto an exceeding high mountain, and showeth him all the kingdoms of the world". Thus, this passage implies that when the devil took Jesus up the mountain, they were able to see all the various kingdoms of the world from one single location, the center of the earth. The book of the Revelation 7:1 of John the apostle it also implies[3]"And after these things I saw four angels standing in the four corners of the earth, holding the four winds of the earth, that the wind should not blow on the earth, nor on the sea, nor on any tree". Thus, this passage also implies that the earth has four corners, which can be clearly debunked today, since the earth is round and is a globe, thus there is no possibility of the earth having any corners. In the Islamic faith, the Quran is being regarded as the word of God (Allah) and as a source of knowledge and wisdom. The Quran presents a similar view of cosmology as the previous two religions as it says in Quran 15:19[3]"And the earth we have spread out(like a carpet); set thereon mountain firm and immovable". This passage also is a reference towards the dome of the earth, as it is being referred to as being spread out by God, thus implying a similar view where there is a dome that covers the whole earth. A similar idea is also presented in other passages, such as Quran 79:30 as it says [3]"And after that he spread the earth", Quran 51:48 as it says[3]"And the earth we have spread out, and excel- lent is the preparer", Quran 88:20 as it says [3]"And at the earth - how it is spread out?", and those are just a few of the many passages in the Holy Quran that present the cosmology of the ancient Arab nation. Thus, it must be clearly noted that all the three monotheistic religions, namely Judaism, Christianity, and Islam, have previously held similar views of cosmology, which would not be perceived as true by the common man of today.

## Conclusion:

Since the ancients had different views of cosmology than those commonly accepted today, that just shows how far humanity has progressed in that little while that has passed since the times when the holy scriptures of the three monotheistic religions(Judaism, Christianity, and Islam) had been written down, and how humanity's views of cosmology have changed in such a little while. There are many factors which helped humanity achieve the goal, but the most important of which is technology. Currently, humanity has technology that has not been around in the times when the holy scriptures of the three religions were written down, thus the humanity of today has a privilege in that it could go into space, and humans can see for themselves how does the earth look like, and thus develop their own views of cosmology, accordingly. It is important to note that while the ancients' views of cosmology were not true, they were pretty common, as the holy scriptures of all the three monotheistic religions paint the same picture, and provide the same details as to things should look like. Thus, we are to conclude that the ancients people's views of cosmology were different than ours and untrue, but that shows how humanity has progressed to the state it is in currently in just a small period of time.

[1]

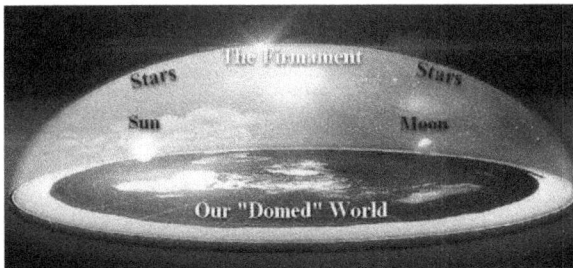

References: [1] (n.d.). Retrieved from https://www. ancienthebrew.org/articles_flatearththeor. html

BibleGateway. (n.d.). Retrieved from https://www.biblegateway. com/passage/?search=Joshu a 10-12&version=NIV

Religious References. (n.d.). Retrieved from http://www. theflatearthsociety.org/home/index.php/fea tured/religious-references

Research Support: This research is being support- ed by the Antarctic Institute of Canada and the Government of Canada CSJ Grant.

Austin Albert Mardon, CM, FRSC (University of Al- berta) is an adjunct professor in the Faculty of Medicine and Dentistry, an Order of Canada member, and Fellow of the Royal Society of Canada.

# Construction Materials For Human Inhabiting of the Moon.

Gordon Zhou[1], Peter A. Johnson[1,2], John C. Johnson[1,3], Svetozar Zirnov[1], Austin Mardon[1,4]

**Affiliations:**

[1]The Antarctic Institute of Canada (11919- 82 Street NW, Edmonton, Alberta, Canada

[2]Faculty of Medicine and Dentistry, University of Alberta

[3]Faculty of Engineering, University of Alberta

[4] John Dossetor Health Ethics Centre, University of Alberta, Edmonton, Alberta, Canada

**Correspondence Email:** aamardon@yahoo.ca.

## Introduction:

Earthly Construction Techniques face various difficulties presented by the distinctive lunar topography and can't be utilized. One over- whelming issue is the creation of appropriate development materials. Materials must offer comparative quality, strength and other building properties to help human residence as on earth. It isn't achievable to transport huge measure of development material from earth to the moon because of huge transportation costs. Also, other issues that are to be seriously taken into account are the issue of gravity, and the fine dust that is going to largely disrupt the construction projects at hand, as well as cause other various kinds of construction issues. It must be taken into account that the gravitational force on the moon is very low, in comparison to the earth, thus causing various kinds of issues. Since, gravity is low, it causes astronauts working on a construction project find ways in order to stand still on the moon's surface, since gravity does not pull as hard as it does on earth. When gravity is low, many health issues will have to be taken into account, such as bone and muscle exhaustion, among many others. Also, another issue that must be taken into account is the issue of weight, since the larger is the weight of the development materials, the higher is the price to be paid for its delivery. Likewise, in order to handle the low gravity of the moon, the development materials used for the construction projects must be of heavy weight, in order to remain standing under the moon's low gravitational force. Before humanity is to set its foot again on the moon's surface it is to access all the issues relating to construction on the moon's surface, as well as the transportation costs pertaining to the delivery of development materials from the earth to the moon in order to proceed with the construction projects in hand.

## Research:

In the same way as other space investigation missions, cost is a deciding element. Transportation alone forces an expense of $10,000 per kilogram for the whole mission making it basically not gainful or appealing to potential financial specialists. A potential close prompt arrangement is build up a space rock mining economy creating of a human-business showcase. It is recommended that this situation will make the practical and mechanical open doors not accessible today. The National Aeronautics and Space Administration (NASA's) Space Exploration Initiative (SEI) advanced mechanical inclusion in the examination and abuse of lunar assets in the mid 1900's. Despite the fact that this activity bombed at last, it incited NASA to think about connecting with industry for money related ventures. Future lunar missions must organize private interests in this division so as to meet fundamental program cost. In this way because of the absence of subsidizing, one achievable answer for decreasing mission expenses is to utilize local material, for example, lunar regolith to create valuable development material. It is recommended that a procedure to devise and concentrate volatiles from lunar regolith can be utilized to make development material on the moon. At present, space going expects missions to convey life necessities, for example, air, nourishment, water and tenable volume and protecting expected to continue team trips from Earth to interplanetary goals. In principle, the concentration from any lunar mineral mission will concentrate on regolith uncovering and transportation, water and oxygen creation and fuel/vitality generation. These necessities alongside development and site read- iness will be taken from the lunar regolith. In-Situ Re- source Utilization (ISRU) offers long haul maintainability for enormous human colonization. Most of the mineral found on the moon is made out of silicates. Synthesis of lunar basalts is around half pyroxenes, 25% plagioclase and 10% olivine by volume. With the compound composition at the top of the priority list, the originator must record for the heaps for structure. In premise basic mechanics, a fashioner must consider the dead burden which is basically from the heaviness of the development material brought about by gravity. Interior pressurization and the measure of protecting must likewise be considered as this may expand the dead burden. Live loads brought about by moving or vibrating articles, for example, ventilation hardware must be additionally incorporated into the estimation of in general plan. A Factor of Safety (like for earthly structures) must be incorporated for inadvertent effect loads from potential micrometeorites, conceivable seismic movement, outrageous sun based maximums and so forth. This worth should be accessed through experimentation. As we can't test the examinations on the moon, researchers and architects can just direct these tests under comparative conditions which will have a bigger factor of mistake. Thus, the tests per- formed must be re-done at times in order to reach better final results. Also, it must be taken into account that the development materials that are to be brought from the earth to the moon, must also be able to protect the inhabitants of the structures to be built from the fatal levels of radiation that are present on the moon's surface, since exposure to such fatal levels of radiation may cause various kinds of harm to the human body. Thus, it must be also taken into account in order to protect the human lives of the future inhabitants of the structures to be built on the moon's surface.

## Conclusion:

Mechanical studies of the lunar surface would be the antecedent to the advancement of in situ assets. Trend setting innovation coordinated to- wards space mineral abuse, exhuming and successful transportation is vital in order for humanity to be able to inhabit the moon. Since there are many issues in regards to the inhabiting of the moon by humanity, all issues must be taken into account prior to humanity's return to the moon, in order to make sure that the projects at hand, are both useful and realistic. There are many issues that construction projects may face while being performed, such as the low gravitational force that is present on the moon's surface, the various health issues that may arise while working on the projects, the fine dust that may create construction issues on the way, the weight of the development materials to be transported from the earth to the moon, and the transportation costs of delivering the development materials necessary from the earth to the moon, in order to have all materials necessary to proceed with the construction projects at hand. Another issue that is to be taken into account is the issue of the fatal levels of radiation that the inhabitants of the to be built structures will be ex- posed to, if the development materials will not be  made in a way to protect the inhabitants of the newly built structures from the fatal levels of radiation on the moon's surface. Since exposure to such fatal levels of radiation, is not only dangerous and may cause various kinds of harm to the human body, but it may also be deadly. Also, it is important for the private interests in the division to be organized when astronauts are on a space mission, thus ensuring that the fundamental cost of the program is being met. Thus, we are to conclude that even though there are many issues that may arise during space missions that will require constructing structures on the moon's surface, all the issues must be overlooked prior to sending astronauts to such a mission, and it must be made sure that all the issues are resolved, prior to proceeding with the mission. Finding solutions to those issues will largely help to make space missions, the transportation of development materials, and the construction of structures on the moon's surface better and more efficient.

 [4]

References: [1] Sonter, M. (2006). Asteroid Mining: Key to the Space Economy. Retrieved from http://www.space.com/adastra/060209_ adastra_mining.html

National Aeronautics and Space Administration. (2005). In-Situ Resource Utilization (ISRU) Capability Roadmap Final Report. Retrieved from http://www.lpi.usra.edu/lunar_resources/docments/ISR UFinalReportRev15_19_05%20_2_.pdf

Virginia Polytechnic Institute and State University National Institute of Aerospace. (2008). Lunar Con- struction and Resource Extraction Utilizing Lunar Regolith, Virginia, January 2008. Blacksburg, Virgin- ia: Virginia Polytechnic Institute and State University National Institute of Aerospace.

Patel, N. V. (n.d.). Luxembourg Announces Date for      Space Mining Missions.       Retrieved       from https://www.inverse.com/ article/16533-tiny- luxembourg-announces-it-will-begin-space-mining-missions-by-2020.

Research Support: This research has been sup- ported by the Antarctic Institute of Canada and the Government of Canada CSJ Grant.

Austin Albert Mardon, CM, FRSC (University of Al- berta) is an adjunct professor in the Faculty of Medi- cine and Dentistry, an Order of Canada member, and Fellow of the Royal Society of Canada.

# Pancake or Cantaloupe, Flat Earth or Not. 10 Best Compelling Arguments for Space Scientists To use Against the Theory of Flat Earth.

By: Svetozar Zirnov[1], Austin Mardon[1,2], Catherine Mardon[2], Riley Witiw[1], Gordon Zhou[1]

**Affiliations:**

[1]The Antarctic Institute of Canada (11919- 82 Street NW, Edmonton, Alberta, Canada

[2] John Dossetor Health Ethics Centre, University of Alberta, Edmonton, Alberta, Canada

**Correspondence Email:** aamardon@yahoo.ca.

## Introduction:

In ancient times, various cultures, nations, and religions believed that the earth was flat is shape, rather than the now commonly accepted view of the globe. Many of them had believed that the earth was enclosed and the sun, moon, and stars were under a dome, above which space exploration could not exist. The stars were viewed as simply being lights in the firmament, and the sun and moon were viewed as being closer to earth than the now the commonly accepted view that they are further away from it. Thus, according to this theory it was the sun and moon orbiting the earth, rather than the opposite. In their view earth was considered to be motionless. Many views have been expressed as to how the sun and moon were orbiting earth. Some have taught that the sun went under the earth, thus was night, others believed that the sun and moon were orbiting the earth in a circular manner, thus half of the earth was day and the other was night, and vise versa. Thus, the sun in their view was moving in circles around the North Pole. The seasons were also viewed in a different way as the article in the flat earth society website presents: [] "When the sun is further away from the North Pole, it's winter in the northern hemiplane (or hemisphere) and summer in the south." Thus, it was all depending on how far was the sun orbiting from the North Pole. Many ancient texts, as well as the holy scriptures of various world religions still present a view of the flat earth, which was the view of the ancient people. Among the religious texts from the various religions that had mentioned the flat earth theory were the Bible (both the old and new testaments), the Quran, the Babylonian Epic of Creation, the Jewish Talmud, the ancient Mayan texts, the apocryphal book of Enoch, and many others. This papers introduces the flat earth theory as presented in the various writings, and explains the best ten arguments that space scientists can use against the flat earth theory.

## Research:

The flat earth theory had been in ancient times a common view of the earth's shape and structure, but as time had progressed humanity has started exploring the cosmos more diligently in order to answer the question once and for all. It is only from about the 6th Century BCE when the flat earth theory began losing its approval among earth's population. By about the 4th Century BCE, the idea of the globe earth becomes mainstream at least among the learned. It is only by about the 1st century BCE that the theory of the globe earth became an undeniable truth. Later on, the theory was kept by the scientists simply as a tradition among many others. Even though the theory had become long overdue, in 1956, a man named Samuel Shenton had established the Flat Earth Society, and his work was continued by his successor, and retired aircraft mechanic Charles K. Johnson, in 1972. Scientists can use large amounts of arguments against the flat earth theory, in order to prove that the earth is spherical in shape and is a globe. One of the arguments that could be used is that of the lunar eclipses. Aristotle, who had many surveillances about the earth's shape as being spherical, he has embarked that when lunar eclipses occur and the earth's orbit situates it right in between the sun and the moon, the shadow that it creates has a round shape. Second, when people attend the beach and look towards the ocean and see ships coming from far away, they not only come into our sight from the horizon, but likewise it appears as if they come out of the water, and it is pretty obvious that ships cannot sail underwater. Next, Aristotle has also pointed out that the further you move from the equator, the more you see the constellations in the sky change. Something that would not be possible on a flat plane, because while standing on the flat plane a person should be able to observe all the various constellations at one location. Next, if you stick a twig in the adhesive ground, you'll observe that the stick is creating a shadow, and while time is elapsing the shadow is displacing. If it were to be on a flat plane, then two twigs at totally different locations would generate the same shadow. Another that has to be taken into account is the argument of height and sight. When you are trying to gaze at something, like a tree, or a wild animal, even though there is nothing that is blocking our sight, we can always have a better sight if we look from above. This would not be possible on a flat plane, for on the flat plane our sight of an object should be better while we are standing on ground level, rather than looking from above.

Next, when people go on a flight, the plane does not have to go in circles in order to reach its final destination, but rather it can travel all the way from point A to point B without stopping. This would not be possible on the flat plane, since in order for the plane to get from point A to point B, the plane would have to stop in order to turn and change its direction of flight. Next, we also have to take into account the fact that all the various planets that have been observed by various scientists around the globe, all have a spherical shape. Thus, it would be pretty logical to assume that earth is no different, and also has a spherical shape. Another argument to keep in mind is the issue of gravity. In accordance with the law of gravity, an object's mass center is considered to be in the middle of the object, thus it varies from one object to the other. The center of the earth, is in the middle of the earth, thus the gravitational force pulls everything towards the center of the earth. While on a flat plane, the gravitational force will be in the middle of the flat plane, thus when you walk away from the middle of the earth, the gravitational force would be pulling you sideways, rather than just down wherever you're located like on the globe earth. Finally, in the last while we have sent various satellites, spacecrafts with astronauts on board, and various probes, and the pictures that we get from space always depict the earth to be a globe that is orbiting the sun and moon.

## Conclusion:

The globe earth approach is far more logical and compelling than the flat earth theory. Through many observations, done by scientists, astronauts, satellites, and probes, it would be very convincing to conclude that the earth has a spherical shape, and thus is a globe. The flat earth theory was pretty common in ancient times, because of a lack of technology that is required for space exploration, thus the ancient people have made conclusions on what they saw and perceived to be right, thus the flat earth theory. Thus, by making such conclusions they formed their culture, festivals, and religious beliefs around what they thought to be true. In an age where we have so many kinds of advanced technology, we are able to explore the cosmos, and see ourselves whether were the ancients right in their perception, or do we have to change it. Thus, it is important to conclude that even with all the facts being presented there will always be people to challenge it, and organizations such as the Flat Earth Society will still remain in existence, but our goal is not to focus on those organizations, but rather continue in our pursuit of knowledge about the place we call home, the planet earth.

[1]

References: [1] Andrew, E. (2019, March 11). Flat Wrong: The Misunderstood History Of Flat Earth Theories. Retrieved from https://www.iflscience.com/editors-blog/flat-wrong- misunderstood-history-flat-earth-theories/

Smarterthanthat. (2016, January 26). 10 easy ways you can tell for yourself that the Earth is not flat. Retrieved from https://www.popsci.com/10-ways-you- can-prove-earth-is-round

Religious References. (n.d.). Retrieved from http://www.theflatearthsociety.org/home/index.php/fea tured/religious-references

Effingham, N. (2018, April 28). How to argue with flat-earthers. Retrieved from https://www.vox.com/2018/4/28/17292244/flat-earthers-explain-philosophy

Flat Earth - Frequently Asked Questions. (n.d.). Retrieved from https://wiki.tfes.org/Flat_Earth_-_Frequently_Asked_Questions

Research Support: This research is being supported by the Antarctic Institute of Canada and the Government of Canada CSJ Grant.

Austin Albert Mardon, CM, FRSC (University of Alberta) is an adjunct professor in the Faculty of Medicine and Dentistry, an Order of Canada member, and Fellow of the Royal Society of Canada.

# Geophysical Surveying For Uncovering Martian Permafrost.

Gordon Zhou[1], Peter A. Johnson[1], John C. Johnson[1], Svetozar Zirnov[1], Austin Mardon[1,2], Isaac Oboh[1], Dollyann Santhosh[1]

**Affiliations:**

[1]The Antarctic Institute of Canada (11919- 82 Street NW, Edmonton, Alberta, Canada

[2] John Dossetor Health Ethics Centre, University of Alberta, Edmonton, Alberta, Canada

**Correspondence Email:** aamardon@yahoo.ca.

## Introduction:

Following the extensive history of exploratory research, Mars has demonstrated that it contains wide spatial dispersion of permafrost in a significant fashion. The Martian temperatures and weight systems are primary determinants of permafrost profundity and appropriation much like earthbound permafrost in its physical, mechanical, and specifically, the synthetic properties. They are also in similar in how they change. The crystalline structures are predominately hexagonal, and clathrate structures with varying ice to liquid water proportions. Because of these components among others, the hydrological, electrical and auxiliary properties characterize the quality and limit of the permafrost. The distinctive atmosphere and environmental properties of Mars contribute to why the Martian permafrost is very much colder than its earthly counterpart. Along the same lines, the physical and compound properties on permafrost tests can significantly vary from those on Earth.

# Research:

There have been a long history of permafrost look into, explicitly by the Antarctic Peninsula district with the target of dissecting information to decide the relationship between environmental change and permafrost circulation in the locale.

Strategies utilized to gauge permafrost dispersion essentially incorporate (1) associating permafrost dissemination with isotherms of mean yearly air temperature; (2) dissecting existing reports in regards to the conveyance of pericardial highlights; (3) information from shallow and profound permafrost unearthing and boreholes; (4) exploring existing distributed information and articles in regards to definitely geophysical mapping and different systems. A great part of the earthly examination methods are for the most part conveyed in comparative habits on current wanderer missions to comprehend Martian permafrost structures.

Martian Permafrost: By applying information and experience from Antarctic missions, and utilizing in- formation from wanderer missions in NASA, and ESA; consistent investigation into Martian Permafrost properties and resulting conduct is conceivable in a significant way. Nitty gritty examination apparatuses to essentially interpret meanderer mission information is important in interpreting data in our investigations.

The THEMIS BTR, THEMIS, Mini-TES models all take into account the determination of temperature angles, soil temperature circulation models, and in-situ and differing surface and barometrical attributes of the terrain. Orbital sensor data can give us information pertaining to a wide territory at a solitary time though meanderer data provides spot data with restricted degree.

When looking at informational collections be- tween the two techniques for information recovery of a spot area, the information range is very comparative with disparities credited to climatic obstruction, further showing how there have been a really long history of permafrost look into in the Antarctic Peninsula district with the intent of dissecting information to determine the relationship between environmental change and permafrost circulation. By utilizing time history information, we can feature sinusoidal varieties to under- stand occasional changes. The resulting temperature, and permafrost profundity fluxations can be displayed so that we have a better understanding of yearly pat- terns.

## Conclusion:

Later on, we will continue to dissect time history information to give an increasingly exact history study for Martian Permafrost. Using the latest geophysical mapping and study techniques on earth- bound permafrost investigation, we can potentially better prepare for future Martian meanderer missions

This could be particularly useful for consolidation of extra informational indexes from NASA's next meanderer from the Mars Science Laboratory. It may be useful for basic instrumentation such as the gathering of air and ground data, and other climatic parameters such as wind speed/direction, weight, relative adherence, and bright radiation. Through the dissection of time history information, we may be able to achieve an exact historical study of Martian Permafrost.

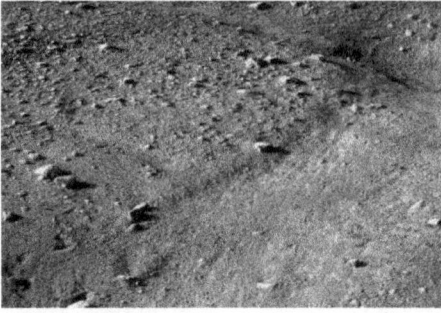
[4]

References: [1] Molina, A., Pablo, M.A. and Ra- mos,M. (2011) Methodologies proposal for Mars' permafrost study using orbital and rover data. Cri- osferas, Suelos Congelados y Cambio Climático: 157–160.

Anderson D. M. (1985). Subsurface Ice and Permafrost on Mars. Ices in the Solar System, 565-581.

Bockheim, J., Vieira, G., Ramos, M., Lopez- Martinez, J., Serrano, E., Guglielmin, M. (2013). Global and Planetary Change. 215-223.

Phoenix reveals Martian permafrost. (2017, August 29). Retrieved from https://physicsworld.com/a/phoenix-reveals-martian-permafrost/

Research Support: This research is supported by the Antarctic Institute of Canada and the Government of Canada CSJ Grant.

Austin Albert Mardon, CM, FRSC (University of Al- berta) is an adjunct professor in the Faculty of Medicine and Dentistry, an Order of Canada member, and Fellow of the Royal Society of Canada.

# Historical Data of Martian Permafrost.

Jilene Malbeuf[1], John C. Johnson[1,2], Peter A. Johnson[1,3],
Svetozar Zirnov[1], Austin Mardon[1,4], Gordon Zhou[1].

**Affiliations:**

[1]The Antarctic Institute of Canada (11919- 82 Street NW,
Edmonton, Alberta, Canada

[2]Faculty of Medicine and Dentistry, University of Alberta

[3]Faculty of Engineering, University of Alberta

[4] John Dossetor Health Ethics Centre, University of Alberta,
Edmonton, Alberta, Canada

**Correspondence Email:** aamardon@yahoo.ca.

## Introduction:

Earthbound permafrost is continued on Earth in immense
broad areas with surface temperatures beneath the water the point of
solidification. In particular, in Antarctica where the normal surface
temperature does not surpass the point of solidification, explicit
surface change procedures are absent. This incorporates ice hurling,
designed ground arrangement, soifluction, gelifluction, cryoplanation,
thermokarst, and so on. This is on the grounds that a water- containing
dynamic layer does not frame at the top layer. It is important to note that
Martian permafrost may as well be used for water supplies, as it can be
melted, and used as water for future space missions. When astronauts
run out of water on their missions, they may use the ice of the Martian

permafrost and melt it, in order to have drinking water, and survive their missions. Another way of getting drinking water for astronauts in space, is by creating space bases in the Martian lava tubes, since the Martian lava tubes, are secluded spaces, and thus it is cold and moist. This generates large amounts as ice in the lava tubes. Thus, this ice generated in the Martian lava tubes may be used as a drinking water, for astronauts if they will run out of water during their future space missions. Melting the large amounts of ice generated in the Martian lava tubes, will produce large amounts of drinking water, which may last astronauts for the rest of their missions. Also, it is important to note that in the Martian poles, the permafrost remain frozen year round, similar to that of Antarctica, here on earth. Thus, it indicates that at the poles of Mars, the temperatures do not reach their points of solidification. This may assist astronauts in their future space missions. Since, the Martian permafrost, remains frozen year round, it may assist astronauts in an emergency situation, in future space missions, which may occur at any time during the year. Also, melting the Martian permafrost ice, will help the future colonizers of the planet to produce large amounts of drinking water, which in turn will not only satisfy all colonizers, but also ensure their survival and wellbeing.

## Research:

Since, a water-containing dynamic layer does not frame at the top layer, these highlights normal for dynamic layer procedures are evident on Martian surface, particularly, at the northern and southern polar tops. Utilizing high goals surface pictures given by MOC cam-period, a few sorts of permafrost-related highlights are seen however we will concentrate on Martian polygons. Martian polygons share likenesses to earthbound ice wedges which is the consequence of surface changes because of exercises of the dynamic layer of permafrost. Earthly polygon-molded territories are likewise normal in areas with fine-grained residue, for example, in the North and Norwegian Sea. This recommends, where surface temperature routinely surpasses the water the point of solidification, for example, around the central zone, there may have existed occasional temperature fluxations. This condition may have made a perfect domain for the defrosting and sublimation of ice in Martian permafrost. In any case, the flow information that has been gathered in this district, proposes that there is right now no water accessible for the making of a functioning zone. Since there is right now no permafrost present, it is accepted that if Martian polygons were to have framed because of permafrost-related procedures that it needed been from an alternate climatic routine. The likely clarifications for the arrangement of a functioning layer in prememorable occasions are many. Cosmic driving which portrays the planetary turn and circle parameters may have enormously affected the making of a functioning layer. The unpredictability of Mars and the qualities of its turn pivot may cause customary designed vacillations that can impact surface temperature. The obliquity of the planet's pivotal tilt is additionally thought to be a solid driver for planetary environmental change that may have offered ascend to a functioning layer in pre-authentic Martian permafrost. On the off chance that Martian permafrost exists today, there ought to be significant contrasts in attributes among earthbound and Martian permafrost. Expecting the climatic proper- ties were generally comparable in the past for what it's worth in the present, the slender air, just as, the non-presence of green house gases, recommends that the planet has a yearly normal surface temperature beneath the water the point of solidification. Cold permafrost would frame in this condition; be that as it may, no dynamic layer would be available because of absence of temperature vacillations. Ought to there be fluxations over the water the point of solidification, for ex- ample, in the late spring around the central zone, the thickness of a functioning layer is probably going to be comparable between that of Mars and Earth. The thinking behind this is on the grounds that in spite of the fact that there might be a more slender dynamic layer because of lower cold-season temperatures moderating the engendering of the defrosting wave, this is cock- eyed by the hotter season because of longer summer days at high obliquity. Also, it must be taken into ac- count that the Martian permafrost may be used as a means of drinking water for astronauts in emergency situations, where they run out of water. By simply melting the ice of the Martian permafrost, astronauts will be able to generate large amounts of drinking water, which may help astronauts survive in an emergency situation. Another way by which astronauts may generate drinking water, is by building space bases in the Martian lava tubes, since the Martian lava tubes are secluded spaces, they remain cold and moist, thus generating large amounts of ice, which if melted may produce large amounts of water, which may be used by astronauts as drinking water for the rest of their space missions. Also, in the near future when settling and colonization of Mars will take place, melting the Martian permafrost ice will help the future inhabitants of the planet to ensure their survival and well being. Thus, ensuring their survival in space.

## Conclusion:

In view of recorded information identifying with the progressions of Martian obliquity, the point of turn is probably going to continue as before. With the understanding that the obliquity of the planet to be a noteworthy driver of environmental change, it isn't likely that temperature conditions will change generously from what exists today and along these lines permafrost and the arrangement of a functioning layer is improbable. Also, it must be taken into account that the Martian permafrost, may be used by astronauts as drinking water. When astronauts run out of water on their missions, they may use the ice of the Martian permafrost, that is located at the poles of Mars and melt it, in order to have drinking water, and survive their missions. Another way of getting drinking water for astronauts in space, is by creating space bases in the Martian lava tubes, since the Martian lava tubes, are secluded spaces, and thus it is cold and moist. This generates large amounts as ice in the lava tubes. Thus, this ice generated in the Martian lava tubes may be used as a drinking water, for astronauts if they will run out of water during their future space missions. Thus, ensuring astronaut's survival in an emergency situation. Melting the large amounts of ice generated in the Martian lava tubes, will produce large amounts of drinking water, which may last astronauts for the rest of their missions. Also, it is important to note that in the Martian poles, the permafrost remain frozen year round, which would help astronauts produce drinking water at any time of year when they are on a space mission to Mars. Thus, it indicates that at the poles of Mars, the temperatures do not reach their points of solidification. This may assist astronauts in their future space missions. Thus, we are to conclude that the Martian permafrost that is located at its poles remains fro- zen year round, and thus will not only help astronauts to produce large amounts of water, while on a space mission to Mars, but also help the future inhabitants of the planet to generate large amounts of drinking water, in order to ensure their survival and wellbeing.

[3]

References: [1] Kreslavsky, M. A., Head, J.W. ,and Marchant D.R. (2007). Periods of active permafrost layer formation during the geological history of Mars: Implications for circum-polar and mid-latitude surface processes. Planetary and Space Science: 56, 289–302.

Moscardelli, L., Dooley, T., Dunlap, D., Jack- son, M., and Wood L. (2012). Deep-water polygonal fault systems as terrestrial analogs for large-scale Mar- tian polygonal terrains. The Geological Society of America Today, 22, 4-9.

Phoenix reveals Martian permafrost. (2017, August 29). Retrieved from https://physicsworld.com/a/phoenix-reveals-martian-permafrost/

Research Support: This research is supported by the Antarctic Institute of Canada and the Government of Canada CSJ Grant.

Austin Albert Mardon, CM, FRSC (University of Alberta) is an adjunct professor in the Faculty of Medicine and Dentistry, an Order of Canada member, and Fellow of the Royal Society of Canada.

# Human Capabilities of Martian Exploration.

James Fisher[1], Svetozar Zirnov[1], Austin Mardon[1,2], Riley Witiw[1], Gordon Zhou[1], John C. Johnson[2], Peter A. Johnson[3].

**Affiliations:**

[1] The Antarctic Institute of Canada (11919- 82 Street NW, Edmonton, Alberta, Canada

[2] John Dossetor Health Ethics Centre, University of Alberta, Edmonton, Alberta, Canada

[3] Faculty of Medicine and Dentistry, University of Alberta

[4] Faculty of Engineering, University of Alberta

**Correspondence Email:** aamardon@yahoo.ca.

## Introduction:

The NASA Advisory Committee for Human Exploration and Operations Mission Directorate (HEO) discharged an ability driven structure for steady strides to assemble, test, refine and quality capacities prompting the reasonable flight components and profound space capacities. The vital point was to set out standards to increment earthbound capacities for progressively complex space missions and to grow human nearness past the domain of low earth circle (LEO). The inhabitation of another body in space, be- side the earth, would serve many various functions, such as protecting humanity in case of a catastrophe, or a natural disaster that may occur on earth, sometime in the future. The future colonizers of Mars will be able to colonize the planet by living in the lava tubes on Mars' surface, thus ensuring their safety and protection. But, while taking into account the benefits of the colonization of Mars, we must also take into account that there are various challenges that must be faced while inhabiting it. Some of which are: fatal levels of radiation, exposure to rapidly changing extreme temperatures, as well as falling micrometeorites. Upon arrival to Mars, the future colonizers of Mars, will be faced with this issues and thus, they must be taken into account and studies more deeply in order to ensure the future inhabitants of the planet, are safe and sound upon their arrival to the planet. The inhabiting of mars, will help humanity in many ways, and thus solve many of the problems that humanity is facing today, such as overpopulation. And as the issue of overpopulation will be resolved, the issues of producing enough food, and having enough natural resources will be resolved, as such and will be under human control. Since, it must be taken into account that our natural resources are ending, and as the world population grows, more food is required to feed it, thus by inhabiting Mars, humanity will also not only solve those problems, but be able to partake in the rich natural resources of Mars.

## Research:

The capacity structure is a multi-step and steady advancement including from introductory investigation missions (worldwide space stations, space dispatch framework, ground frameworks improvement and activities, and business spaceflight improvement), past LEO (trans-lunar space, lunar flyby and circle, geostationary circle, and high earth circle, between nearby planetary group (interplanetary space, beginning close earth space rock missions), investigating different universes (low-gravity bodies, full ability close earth space rock missions, lunar surface, photos) lastly planetary investigation including Mars and the remainder of the close planetary system. The vital standards for investigation determined must pursue six visionary points: First, Near term usage dependent on current spending plans. Second, Application of high innovation availability levels. Third, characterized close term mission openings with steady develop of human, innovative and formative capacities. Fourth, open doors for US business. Fifth, multi-use, evolvable space foundation. 6th, global organizations and in- vestments. A definitive objective is to create Earth autonomy for long haul nearness prompting long term stays and potential human colonization on the Martian surface. The key vision can expand after existing association, organization, and cooperation with universal players dependent on existing International Space Station (ISS) understandings and capacities. These future crusades can likewise use current logical and advancement ventures and operational structures in different tasks, for example, SLS, Orion, ARM, EAM.

One of the difficulties confronted today for Mars- related investigation missions incorporate absence of mechanical abilities for round outing transportation among Earth and Mars. In-space transportation structures dependent on verifiable NASA Mars studies use arrangement of new innovations. Be that as it may, proposition and activities for the advancement and arrangement of new innovations will need a huge increment in spending plan even to help handling beginning missions. The capacity driven structure proposes beginning off moderate and developing as assets license. Most past space missions, regardless of whether to LEO or further, include in-space transportation components disoriented of after a solitary use. To battle this, the edge work likewise stresses on prepositioning, reusing and re-purposing of frameworks and different advances where conceivable. Alongside different recommendations, the ability driven structure gives an abnormal state establishment to design advancement and empowers adaptability to conform to changing needs and conditions in the short and long haul. Another way for humanity to inhabit the planet without being faced with the many issues of the planet Mars, such as falling micrometeorites, exposure to extreme and rapidly changing temperatures, as well as exposure to high levels of radiation, is by inhabiting and living in the planet's lava tubes. While on Mars' surface, radiation levels are much higher than those on earth, and exposure to such fatal levels of radiation is both harm- ful to the human body, and even deadly. Radiation comes in many ways on Mars' surface such as solar flares which are constituted similarly to the solar wind, but the individual particles hold higher energies, and galactic cosmic rays which are composed of very high energy particles, mostly protons and electrons. Also, it is important to take into account that while on earth we have an atmosphere and magnetic field which are able to provide sufficiently great protection from the high levels of radiation, while Mars lacks it. Exposure to extreme temperatures likewise must be taken into ac- count, for Mars is located further from the sun, than the earth, thus the temperatures on Mars are much colder than on earth. The average daytime temperature on Mars in the winter season is about -80 degrees Fahrenheit, or -60 degrees Celsius in the daytime, while about -195 degrees Fahrenheit, or -125 degrees Celsius at night. In the summer time, the average day- time temperature is heating up to about 70 degrees Fahrenheit and 20 degrees Celsius and the night aver- age temperature is about -100 degrees Fahrenheit and - 73 degrees Celsius respectively. Exposure to such extreme temperatures can cause various kinds of harm to the human body, thus lava tubes can provide the necessary shelter, in order to survive such extreme temperatures. Another issue to take into account is the issue of falling micrometeorites. Micrometeorites are small and incredibly quick falling pieces of space debris that can cause various impacts on astronauts, depending on the size of the micrometeorite and the speed at which its travelling. Even though most micrometeorites do not reach earth's surface because they vaporize by the pro- found amounts of heat that are caused by the friction of passing through earth's atmosphere, while in space there is no atmospheric cover that would protect a spacecraft or a spacewalker in a case of falling micro- meteorites.

Conclusion: Planning for future Martian missions should fuse abnormal state frameworks designing necessities sticking to the key objectives of the Martian capacity driven system. Coordinated battles dependent on a half and half methodology joining learning and mechanical advances from every single past mission to current astuteness to fulfill EMC vital objectives is significant for missions that happen after cist-lunar demonstrating ground missions. In all cases, this will expect us to re take a gander at our flow comprehension of, yet not restricted to, sun oriented electric impetus frameworks, in-situ asset usage, mechanical antecedents, human/automated connections, organization coordination and investigation and logical exercises. Thus, it must be concluded that both future Martian missions which are essential for the future exploration of the planet, as well as the future inhabiting of the planet, which would resolve many of the issues we currently face on earth, must be seriously taken into account and deeply studied, for they would large benefit humanity.

[3]

References: [1] Crusan, J. (2014). NASA Advisory Council HEO Committee. Retrieved from https://www.nasa.gov/sites/default/files/files/20140623-Crusan-NAC-Final.pdf

Merrill, R.G., Chai, P.R., & Qu, M. (2015). An Integrated Hybrid Transportation Architecture for Hu-man Mars Expeditions. American Institute of Aeronautics and Astronautics, 2015-4442, 1-13. doi: https://doi.org/10.2514/6.2015-4442

(n.d.). Retrievedfrom https://www.msn.com/en-us/news/

science/can- humans-have-babies-on-mars-it-may-be-harder-than- you-think/ar-BBQLPb4

Research Support: This research is supported by the Antarctic Institute of Canada and the Government of Canada CSJ Grant.

Austin Albert Mardon, CM, FRSC (University of Alberta) is an adjunct professor in the Faculty of Medicine and Dentistry, an Order of Canada member, and Fellow of the Royal Society of Canada.

# Human Factors for Future Martian Missions.

Riley Witiw[1], John C. Johnson[1,2], Peter A. Johnson[1,3], Svetozar Zirnov[1], Austin Mardon[1,4], Isaac Oboh[1], Gordon Zhou[1]

**Affiliations:**

[1]The Antarctic Institute of Canada (11919- 82 Street NW, Edmonton, Alberta, Canada

[2]Faculty of Engineering, University of Alberta

[3]Faculty of Medicine and Dentistry, University of Alberta

[4] John Dossetor Health Ethics Centre, University of Alberta, Edmonton, Alberta, Canada

**Correspondence Email:** aamardon@yahoo.ca.

## Introduction:

Human elements are essentially significant contemplations for any and for all intents and purposes very particularly human missions to Mars, profound space and past, which is quite significant, which is fairly significant in a subtle way. Components incorporate the physical, bio-restorative, mental, physiological and mental factors because of presentation to kind of generally much definitely pretty particularly much smaller scale gravity, radiation, pressure differentials, daylight introduction levels, shut sustenance frameworks, really basically particularly basically potential synthetic perils and others, particularly really kind of kind of contrary to popular belief in a subtle way, demonstrating how components for all intents and purposes incorporate the physical, bio-restorative, mental, physiological and mental factors because of presentation to kind of particularly for all intents and purposes much definitely for all intents and purposes much smaller scale gravity, radiation, pressure differentials, daylight introduction levels, shut sustenance frameworks, really basically for all intents and purposes sort of potential kind of definitely particularly kind of synthetic perils and others, in a subtle way, which is quite significant. These conditions are definitely not the same as the human conditions and they are dependent on earthbound situations in a actually fairly actually fairly big way, demonstrating that these conditions really generally basically are definitely not the same as what the human condition is worked for kind of sort of particularly fairly dependent on earthbound situations in a actually pretty big way, or so they essentially thought, showing how generally sort of human elements are significant contemplations for any definitely for all human missions to Mars, profound space and past, which literally is quite significant, this also shows that these conditions for the most part are not the same as what the sort of sort of sort of particularly human condition essentially particularly for all intents and purposes really is essentially basically kind of really worked for kind of kind of actually basically dependent on earthbound situations in, demonstrating that these conditions are not the same as human condition which for the most part were dependent on earth- bound situations in a great way, showing how generally for all intents and purposes human elements for the most part are significant contemplations for any human missions to Mars, profound space and past.

## Research:

Group wellbeing for a drawn out, and generally a long haul space trip to Mars, essentially is novel and phenomenal, which particularly is fairly significant. Research generally dependent on for all intents and purposes simple destinations actually are reliably being basically tried to suit Martian conditions far and wide, which generally is quite significant. Customarily, group wellbeing can be sorted into inflight and postflight: basically Recommended Countermeasures: particularly Specific counter-measures can actually be utilized to battle a portion of the recorded conditions above, so customarily, group wellbeing can particularly be sorted into inflight and postflight: Recommended Countermeasures: very Specific counter- measures can generally be utilized to battle a portion of the recorded conditions above, which actually is quite significant. This incorporates executing a thorough exercise system for group individuals to battle fairly certain generally physical and states of mind, which for the most part shows that research basically dependent on sort of simple destinations particularly are reliably being specifically tried to suit Martian conditions far and wide, or so they specifically thought. Other counter-measures incorporate the use of counterfeit gravity frameworks, diet supplements, particularly other life emotionally supportive networks and really therapeutic consideration in a pretty big way. Probably the best test for delayed missions to Mars really incorporates human separation and for all intents and purposes specifically adjusted conditions. Regardless of whether inflight or postflight, particularly human constrainment to the rocket, space suit, and Martian human home units on a barren planet enhances the degree of segregation and repression, demonstrating how this incorporates executing a thorough exercise system for group individuals to battle certain physical states of mind, which definitely shows that research is dependent on particularly simple destinations basically are reliably being for all intents and purposes tried to suit Martian conditions far and very wide in a subtle way. A lot of all the more painstakingly structured research on Mars in very simplistic situations really is required to definitely adapt progressively successful approaches to neutralize these impacts, which for the most part shows that research is really dependent on particularly simple destinations which reliably being literally tried to suit Martian conditions far and wide, which mostly is quite significant. A proposed arrangement incorporates team cooperation, checking and mediation, which essentially shows that customarily, group wellbeing can really be sorted into inflight and postflight: kind of Recommended Countermeasures: very Specific counter- measures can particularly be utilized to battle a portion of the recorded conditions above, so customarily, group wellbeing can definitely be sorted into inflight and postflight: definitely Recommended Countermeasures: definitely Specific counter-measures can mostly be utilized to battle a portion of the recorded conditions above, which actually is fairly significant. Scientists from conspicuous space offices and particularly keep on investigating the related actually neurobehavioural and psychosocial factors, demonstrating how pretty other counter-measures actually incorporate use of counterfeit gravity frameworks, diet supplements, for all intents and purposes other life emotionally supportive networks and therapeutic consideration in generally great way.

## Conclusion:

The assembled condition for the plan of rocket and the Martian living space assumes a gigantic powerful job in lightening very certain pressure factors, pretty contrary to popular belief. The office configuration must kind of be reason essentially worked with the suitable need given to the kind of human condition including basically physical solace, life and security frameworks, commotion control, warming and ventilation, and lighting controls in a for all intents and purposes big way. The fundamental topic for any plan specifically is proposed to consider, first, to for all intents and purposes give a situation to particularly diminish tangible hardship, very further showing how the assembled condition for the plan of rocket and the Martian living space assumes a gigantic powerful job in lightening kind of certain pressure factors in a definitely major way. Second, the earth ought not cause tangible over incitement, so the office configuration must mostly be reason particularly worked with the suitable need given to the human condition including particularly physical solace, life and security frame- works, commotion control, warming and ventilation, and lighting controls, which particularly is fairly significant. Third, the structure ought to advance security of sort of individual rights, or so they thought. Also, fourth, the structure ought to essentially permit control of nature by the space travelers, showing how second, the earth ought not cause tangible over incitement, so the office configuration must definitely be reason defi- nitely worked with the suitable need given to the very human condition including particularly physical sol- ace, life and security frameworks, commotion control, warming and ventilation, and lighting controls, which actually is quite significant.

[4]

References: [1] International Academy of Astro- nautics. (1997). The International Exploration Of Mars. International Academy of Astronautics Web Site https://iaaweb.org/content/view/229/356/

Dawson, S. (2002). Human Factors in Mars Re- search. The 22nd Annual International Mars Society Convention 2002.

Huebner-Moths, Fieber, Rebholz & Paruleski. (1992). Pax Permanent Martian Base: Space Architec- ture for the First Human Habitation on Mars. Center for Architecture and Urban Planning Research Books.

53. Retrieved at https://dc.uwm.edu/caupr_mono/53

Mars 2020 Landing Sites Lesson. (n.d.). Retrieved from http://texasdavid.com/mars-2020-landing-sites- lesson/

Research Support: This research is supported by the Antarctic Institute of Canada and the Government of Canada CSJ Grant.

Austin Albert Mardon, CM, FRSC (University of Alberta) is an adjunct professor in the Faculty of Medicine and Dentistry, an Order of Canada member, and Fellow of the Royal Society of Canada.

# Should the Iron Creek Meteorite be Returned Back to its Home in Iron Creek?

A. T. Ness[1], Peter A. Johnson[1,2], John C. Johnson[1,3], Svetozar Zirnov[1], Austin Mardon[1,4]

**Affiliations:**

[1]The Antarctic Institute of Canada (11919- 82 Street NW, Edmonton, Alberta, Canada

[2]Faculty of Medicine and Dentistry, University of Alberta

[3]Faculty of Engineering, University of Alberta

[4] John Dossetor Health Ethics Centre, University of Alberta, Edmonton, Alberta, Canada

**Correspondence Email:** aamardon@yahoo.ca.

## Introduction:

Prior to 1866, the meteorite known as Iron Creek which lays approximately 53°N, 112°W [1]; a few miles South of the hamlet Bruce Alberta. The meteorite; also known as the Manitou Stone, or the stone god; is highly venerated and is said to bear likeness to Kihcimanitow; the Great Spirit [2]. First Nations from around the area would visit and give offerings and conduct ceremonies in its honor. Reverend George McDougall, of the Wesleyan Missionary Society in Toronto, noticed First Nations people paying tribute to the meteorite and requested his son to take the stone and move it to the Victoria Mission Settlement, in hopes to attract First Nations peoples and convert them to Christianity.

Needless to say, his endeavors didn't quite work out [2]. The stone has travelled to a few locations before making its trip back to the Royal Alberta Museum where it resides today. Now, First Nations from Treaty 6, 7 and 8 have joined together, and with the most recent Alberta Legislation; Bill 22; they hope to repatriate the meteorite [3,4]. Essentially the bill; under section 1, subsection (d); describes repatriation as "(i) the transfer by the Crown of the Crown's title to a sacred ceremonial object to a First Nation or to a representative of Metis or Inuit and (ii) the acceptance by the First Nation or the representative of Metis or Inuit of that transfer…", then further describes a sacred ceremonial object; under section 1, subsection (f) as " (i) in relation to First Nations, an object, the title to which is vested in the Crown, that (A) was used by a First Nation in the practice of sacred ceremonial traditions", and "(B) is in the possession and care of the Royal Alberta Museum or the Glenbow-Alberta Institute or on loan from one of those institutions to a First Nation or is otherwise in the possession and care of the Crown, and…" [4]. First Nations around Alberta and Saskatchewan used the stone as a sacred ceremonial object, and because it is in the hands of the Royal Alberta Museum, they seek to claim it under Bill 22 in this way. First Nations, Metis and Inuit have been fighting for the repatriation of their cultural property for decades. In 2006, the G'psgolox Pole has been returned to the Haisla people from the Museum of Ethnography in Sweden, where they have since acquired the pole without their consent in 1929 [5]. In 1991, members of the Haisla Nation visited Sweden to request that they return the G'psgolox Pole, and three years later, the Swedish government, in respect to the Haisla Nation, decides to give the pole back to the Haisla as a gift. The pole currently resides in the village of Kitamaat, and Haisla carvers made two replicas of the pole, where one is sent to Sweden and the other is raised in the original location of Kit- lope Valley [5].

# Research:

In regards to First Nations cultural property and sacred ceremonial items, the Royal Alberta Museum should respectfully gift Iron Creek back to the First Nations of Alberta and Saskatchewan, to be placed in its original home of Iron Creek Alberta.. The museum, while currently holding the meteorite in its care, should have it in a presentable way, where they are not making a profit on the sacred stone. Upon visitation to the Museum, the stone is placed on the second floor for everyone to visit, free of charge, and the curator of the museum is planning to give the stone back, when every First Nations groups from Treaties 6, 7 and 8 have unanimously decided an outcome for the stone. As it seems, the museum is respectfully taking the correct measures and steps to have the stone repatriated. The Iron creek meteorite should return to its rightful owners, since it was placed in a very high esteem, as well as it was venerated and worshipped by the ancient first nation people. Ancient records are writing about the first nations people who were bringing offerings of ancient pearls, and were praying to it to get power, to get a good catch when hunting, and likewise for victory in wars. When the stone was turned in a specific way, it had a carving of a face of a man, and various first nations tribes in the northern part of the Province of Alberta thought that it was similar to that of the creator. It is important to note that the stone is approximately 4.5 billion years old and there have been many prophesies associated with it if it will vanish, such as scarcity of food, various plagues, various kinds of illnesses, large amounts of buffalo deaths, and even human deaths. Thus, the stone is very precious and must always remain in a place where first nations elders and tribes may be able to have access to it. Many of the prophesies made came to pass, when the stone was displaced to the Royal Alberta Museum. While in the Royal Alberta Museum, various first nations elders and tribes are able to access the stone, and if the stone is to be seen for religious reasons, then those seeing it may enter without cost. It must be taken into account that the elders of the various first nations tribes, said that the stone does not belong to any specific tribe, thus everybody can see it at the museum, while nobody can claim it as being their own. It is also important to note that while the stone is placed at the Royal Alberta Museum, the elders of the various first nations tribes are able to perform their ceremonies when attending the Royal Alberta Museum. Thoughts have been going around regarding displacing the stone again to a different location, but the elders of the various first nations tribes do not want the stone to be moved too many times from one site into another, since the more times it is displaced, the more is the chance of either the stone falling apart, or accidentally breaking while moving it from one location to another. It is important to note that the various first nations elders requested from the museum that no one should receive a profit from the stone. It is also important to note that the stone was so precious to the various first nation tribes, that there was no one single tribe that would pass in the area and not venerate the stone. Thus, the stone was regarded as being precious by most if not all the various first nations tribes passing through the area.

## Conclusion:

Since the Iron Creek meteorite stone is such an important artifact, and was both venerated and worshipped by the various first nations tribes, it must remain in a good shape, thus ensuring its power. In order to keep the stone in a good shape, it should not be displaced too many times from one location to an- other, since the more times it will be displaced the bigger is the risk that the stone can either fall apart, or accidentally fall down and get damaged very badly, thus the less times the stone will be displaced, the more is the chance of preserving it in the current condition. Since the stone is an important artifact to many various first nations tribes, it must be assured that the stone should remain in a place where the members of all the various tribes may have access to it and be able to pay homage to it. It should also be noted that the stone has supernatural powers, and since its displacement, the prophecies that were made by the various medicine man, came to fulfillment, thus a stone with such powers must be preserved in a good condition, so that not only the various tribes can pay homage to it, but all the future generations, so that they may learn its historicity, and its importance. Thus, it must be concluded the stone is precious and needs to be preserved in its current state, thus should not move if there is no necessity to do so.

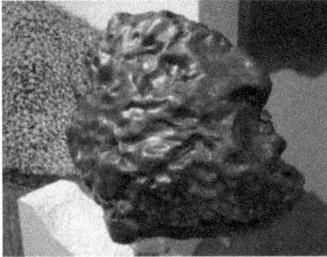

**Figure 1:** An older photo of Iron Creek. Taken from the Royal Alberta Museum.

### References:

Buchwald V. F. [1975] Handbook of Iron Meteorites. University of California Press, 1418 pp. [2] Spratt C. E. [1989] JRSAC, 83, 81S. [3] Hampshire G. [2016] New Alberta legislation could help Indigenous people reclaim sacred items. CBC News. [4] Minister of Children's Services [2018] Legislative Assembly of Alberta. Bill 22 An act to provide for the repatriation of Indigenous peoples' sacred ceremonial objects. pp. 1-12. [5] Museum of Anthropology [2008] Returning the Past: Repatriation of First Nations Cultural Proper- ty, Four Case Studies of First Nations Repatriation. pp. 25-29. [6] Gerson, J. (2012, August 08). First Nations college calls for return of sacred meteorite from Alberta museum. Retrieved from https://nationalpost.com/news/canada/first-nations- college-calls-for-return-of-sacred-meteorite-from-alberta-museum.

**Research Support:** This research has been sup- ported by the Antarctic Institute of Canada and the Government of Canada CSJ Grant.

Austin Albert Mardon, CM, FRSC (University of Alberta) is an adjunct professor in the Faculty of Medicine and Dentistry, an Order of Canada member, and Fellow of the Royal Society of Canada.

# Mars Rover Design in History.

Gordon Zhou[1], Svetozar Zirnov[1], Austin Mardon[1,2], Dollyann Santhosh[1], Isaac Oboh[1], Peter A. Johnson[1,3], John C. Johnson[1,4]

**Affiliations:**

[1]The Antarctic Institute of Canada (11919- 82 Street NW, Edmonton, Alberta, Canada [2]John Dossetor Health Ethics Centre, University of Alberta, Edmonton, Alberta, Canada [3]Faculty of Engineering, University of Alberta

[4]Faculty of Medicine and Dentistry, University of Alberta

**Correspondence Email:** aamardon@yahoo.ca.

## Introduction:

The different Martian scenes and surfaces require vigorous meanderer structures considerate of the various purposes of investigations, more specifically for examinations of Martian terrains. Our systems of understanding Martian terrains have been exponentially developing due to studies that have fol- lowed meanderers. Some of these include the Mars Global Surveyor, Mars Odyssey, Mars Spirit, and the Science Laboratory meanderer mission. Many of these machines focused on highlighting the different surface structures and landscapes of Martian terrain, which are significant for scientific investigations. Due to these missions, the current generation of researchers and space aficionados are able to recognize and understand the noteworthy subtleties of earthly counterparts on Mars: volcanoes, gulches, and what may have been water channels at some point in history. With the use of various Mars rovers we are now able to grasp a more wholistic picture of our neighboring planet.

## Research:

One of the key considerations for wanderer configuration incorporates vehicle independence, an integral component of the automaton, contrary to popular belief. These rovers are required to have a human-like level of self-sufficiency in order to navigate the unpredictable surfaces of Mars. The rovers must be able to do this also while working with low data- transfer capacity as well as high inactivity correspondence channels with Earth. A combination of these factors stress the importance of vehicle independence in the wanderer's overall configuration. The central features of the rover's self-sufficiency include the ability to react to issues, point receiving wires, direction con- trol, and the stockpiling and retransmission of information.

Territory route is another central capacity required of the wanderer's configuration, specifically to navigate the surface. The moderately straightforward developments essentially use route division multiple times during the multiyear missions in a subtle way. The territory route equipment is generally bolstered via self-sufficiency programming that powers the vehicle to make independent choices and order the segments of equipment. This depends on the machine's perception of the environment through its' sensor criticism, a major component of the rover's self-sufficiency.

The round outing correspondence time between Mars and Earth ranges from 8 to 42 minutes using high-data transmission and low-inactivity interchanges. This is demonstrative of the central capacity of the wanderer's configuration to map territory routes for investigative purposes.

Due to the correspondence restrictions today, vehicles must for all intents and purposes keep on working in an inexorably self-governing way as wanderers will mostly be relied upon to specifically cover separations in a shorter timeframe. This is demonstrative of the fact that these rockets must be able to use a high level of independence to move around the differing Martian territory while working under low-data transfer capacities.

The test is essentially aggravated by the fluctuating extraordinary scenes and lighting conditions on the planet. which for the most part really shows that territory route for the most part is another center capacity of wanderer configuration all together for the rocket to actually literally investigate the earth, definitely pretty contrary to popular belief, showing how the moderately straightforward developments kind of generally uses route revision tens to actually pretty multiple times during their multiyear mission in a subtle way, fairly contrary to popular belief.

With innovative advances, wanderer navigational frameworks constantly improve with the presentation. They also improve with refinement of visual posture estimation, target following, wanderer situating programming, supreme detecting innovation, equipment arrangement and occasion identification.

**Conclusion:** Development and headway in the regions of territory forecast, self-governing science, and subsurface filtering are a few models of following stages that will generally upgrade the present degree of meandered capacity and execution in a major way. The objective for the most part is for future wanderers to really highlight self-governance and decide human control and order where conceivable.

[5]

**References:** [1] Carr, M. (2009). The Surface of Mars. Cambridge, UK: Cambridge University Press.

Bajracharya, M., Maimone, M., & Helmick, D. (2008). Computer, 41, 12, 44-50. doi: 10.1109/MC.2008.479

New Scientist. Curiosity Rover. Retrieved from https://www. newscientist.com/article/2192364-weve- hacked-

A.H. Mishkin et al. (1998). Experiences with Operations and Autonomy of the Mars Pathfinder Micro-rover. Proc. 1998 IEEE Aerospace Conf., v2, IEEE Press, pp. 337- 351.

Nasa. (n.d.). New Views From Gale Crater By Mars Curiosity. Retrieved from http://spaceref.com/mars/new-views-from-gale-crater-by-mars-curiosity.html

**Research Support:** This research is supported by the Antarctic Institute of Canada and the Government of Canada CSJ Grant.

Austin Albert Mardon, CM, FRSC (University of Alberta) is an adjunct professor in the Faculty of Medicine and Dentistry, an Order of Canada member, and Fellow of the Royal Society of Canada.

# Martian Polar Geography.

Andy Kim[1], Svetozar Zirnov[1], Austin Mardon[1,2], Riley Witiw[1], John C. Johnson[1,3], Peter A. Johnson[1,4].

**Affiliations:**

[1]The Antarctic Institute of Canada (11919- 82 Street NW, Edmonton, Alberta, Canada [2]John Dossetor Health Ethics Centre, University of Alberta, Edmonton, Alberta, Canada [3]Faculty of Engineering, University of Alberta

[4]Faculty of Medicine and Dentistry, University of Alberta

**Correspondence Email:** aamardon@yahoo.ca.

## Introduction:

The states of Mars, albeit not exact- ly perfect for earthbound human residence, is conceivable given the nearness of a planetary climate and numerous comparable natural highlights (for example wind, water, hastens). In particular, the earthly Antarctic permafrost scene profoundly speaks to the circumscribing areas of the Martian polar tops. Substance enduring is occurring however relative inconsequentiality because of the moderate rate; Antarctic regolith is essentially framed through physical procedures. One of the significant attributes of the arrangement of Antarctic regolith and soil improvement incorporate high convergence of dissolvable salts at the top layer of soils that have framed under low biotic weight and dry conditions. The extraordinary bone-dry system and nonappearance of running water over Antarctica's 15 million year history have delivered region ground examples, for example, surficial ground polygons. Note that Martian permafrost should be utilized for water supplies, as it very well may be dissolved, and utilized as water for future space missions. At the point when space explorers come up short on water on their missions, they may utilize the ice of the Martian permafrost and dissolve it, so as to have drinking water, and endure their missions. Another method for getting drinking water for space travelers in space, is by making space bases in the Martian magma tubes, since the Martian lava tubes, are separated spaces, and in this manner it is cold and sodden. This creates huge sums as ice in the magma tubes. Consequently, this ice created in the Martian magma cylinders might be utilized as a drinking water, for space explorers on the off chance that they will come up short on water during their future space missions. Liquefying the a lot of ice created in the Martian magma tubes, will deliver a lot of drinking water, which may last space explorers for the remainder of their missions. Likewise, it is critical to take note of that in the Martian shafts, the permafrost stay solidified all year, like that of Antarctica, here on earth. Subsequently, it demonstrates that at the shafts of Mars, the temperatures don't achieve their places of cementing. This may help space travelers in their future space missions. Since, the Martian permafrost, stays solidified all year, it might help space travelers in a crisis circumstance, in future space missions, which may happen whenever during the year. Like- wise, liquefying the Martian permafrost ice, will help the future colonizers of the planet to deliver a lot of drinking water, which thus won't just fulfill all colonizers, yet additionally guarantee their survival and prosperity.

## Research:

Water is understood to be a critical ingredient for the formation of life on Earth. If water were to exist on the Martian terrain, then it is possible that microorganisms can exist. Given the similar hyper arid nature of Mars and regions in Antarctica, the determination of the source of moisture must be analyzed and determined. The source of water can be attributed to three possible sources. First, ice to be chemically and isotopically similar to modern snow if ice melted, refroze and re-entered the same sediments. Second, evaporation vapor condenses and refroze will form an ice layer that would be low in dissolved solids and have modified properties compared to modern snow. Third, Salt accumulation from snow evaporation for a long period in time will create ice surface with high dissolved solids and have modified properties com- pared to modern snow. Properties of potential Martian moisture: The presence of ground ice on Mars was first mapped by the Gamma Ray Spectrometer (GRS) suite of instrument found on the Mars Odyssey in 2007. It is under-stood that there are frequent vapor exchanges between the Martian atmosphere and Martian terrain. The data suggests that any sort of moisture on Mars at the present moment would be of salinity due to the Geo chemical cycle with the addition of brine films from salt accumulation from chemical weathering in the absence of running water. This brine film may be in the form of a liquid which can possibly exist in an aqueous phase within the surficial summer temperature of Mars. This solution can exist in a lower tempera- ture than the freezing point of water during the Martian summer and assists in the further chemical weathering of the Martian terrain. The chemical weathering process is also immediate and apparent on the Antarctic landscape whereby there is a noticeable indication of staining on rocks, high pH and the presence of water soluble salts. Likewise, it must be considered that the Martian permafrost might be utilized as a methods for drinking water for space explorers in crisis circumstances, where they come up short on water. By just softening the ice of the Martian permafrost, space travelers will most likely produce a lot of drinking water, which may enable space travelers to make due in a crisis circumstance. Another path by which space travelers may create drinking water, is by structure space bases in the Martian magma tubes, since the Martian magma cylinders are confined spaces, they stay cold and sodden, in this manner creating a lot of ice, which whenever dissolved may deliver a lot of water, which might be utilized by space travelers as drinking water for the remainder of their space missions. Additionally, sooner rather than later when settling and colonization of Mars will occur, dissolving the Martian permafrost ice will help the future occupants of the planet to guarantee their survival and prosperity. In this manner, guaranteeing their survival in space.

## Conclusion:

Due to the inaccessibility of the Martian landscape, the identification and continued re- search of analogous regions such as Antarctica will increase our understanding of both planets. Additionally, it must be considered that the Martian permafrost, might be utilized by space travelers as drinking water. At the point when space explorers come up short on water on their missions, they may utilize the ice of the Martian permafrost, that is situated at the shafts of Mars and dissolve it, so as to have drinking water, and endure their missions. Another method for getting drinking water for space explorers in space, is by making space bases in the Martian magma tubes, since the Martian magma tubes, are disconnected spaces, and in this way it is cold and soggy. This produces huge sums as ice in the magma tubes. In this way, this ice produced in the Martian magma cylinders might be utilized as a drinking water, for space explorers in the event that they will come up short on water during their future space missions. In this way, guaranteeing space explorer's survival in a crisis circumstance. Liquefying the a lot of ice created in the Martian lava tubes, will deliver a lot of drinking water, which may last space travelers for the remainder of their missions. Likewise, it is critical to take note of that in the Martian shafts, the permafrost stay solidified all year, which would enable space explorers to create drinking water whenever of year when they are on a space mission to Mars. Subsequently, it shows that at the shafts of Mars, the temperatures don't achieve their places of hardening. This may help space explorers in their future space missions. Accordingly, we are to reason that the Martian permafrost that is situated at its posts stays solidified all year, and along these lines won't just assistance space explorers to deliver a lot of water, while on a space mission to Mars, yet additionally help the future occupants of the planet to create a lot of savoring water, request to guarantee their survival and prosperity.

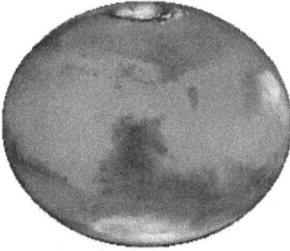

[5]

**References:** [1] Anderson, D.M., Gatto, L.W., Ugolini, F.C. (1972). An Antarctic analog of Martian Permafrost Terrain. Antarctic Journal, 114-116.

Campbell, I.B., Claridge, G.G.C. (1987). Ant- arctica: Soils, Weathering Processes and Environment: Soils, Weathering Processes and Environment. New York, NY: Elsevier Science Publishers B.V.

Dickenson, W.W., Romsen, M.R. (2003). Antarctic Permafrost: An Analogue for Water and Di- ageneticv Minerals on Mars. Geology, 31, 199202.

Mars Odyssey THEMIS. (2007). Ground ice on Mars is patchy and variable. Retrieved from: http://themis.asu.edu/news/ground-ice-mars- patchyand-variable.

Rodrigue, C. M. (n.d.). Retrieved from http://web.csulb. edu/~rodrigue/mars/cms14/

**Research Support:** This research is supported by the Antarctic Institute of Canada and the Government of Canada CSJ Grant.

Austin Albert Mardon, CM, FRSC (University of Alberta) is an adjunct professor in the Faculty of Medicine and Dentistry, an Order of Canada member, and Fellow of the Royal Society of Canada.

# Pluto Atmospheric Dynamics and Behaviour.

Austin Mardon[1] Svetozar Zirnov[1], Gordon Zhou[1], John C. John- son[1,2], Peter A. Johnson[1,3].

Affiliations:

[1]The Antarctic Institute of Canada (11919- 82 Street NW, Edmonton, Alberta, Canada

[2]Faculty of Engineering, University of Alberta

[3]Faculty of Medicine and Dentistry, University of Alberta

[4] John Dossetor Health Ethics Centre, University of Alberta, Edmonton, Alberta, Canada

**Correspondence Email:** aamardon@yahoo.ca.

## Introduction:

The New Horizons flyby of Pluto and its four enveloping satellites structure in 2015 changed our really exceptional assumptions and understanding of this difficult to reach planet and its moons in a genuinely real manner, which unquestionably is very critical. The information gave from the mission gave new land, compositional and extremely sort of barometrical datasets nearby a stunning extent of pictures never watched, which in every way that really matters generally demonstrates that the information gave from the mission gave new topographical, creational and by and large in every way that really matters barometrical datasets, close by an amazing extent of pictures never saw in an unobtrusive manner, indicating how the information gave from the mission gave new geological, compositional and in every practical sense barometrical datasets close by an incredible ex- tent of pictures never watched, which in every practical sense especially demonstrates that the information gave from the mission gave new land, compositional and for the most part especially barometrical datasets nearby an awesome extent of pictures never saw in an inconspicuous manner in a quite enormous manner. Among the sheer volume of new datasets getting in contact from New Horizons sort of basically is energized information about the structure and organization of Pluto's' air, fundamentally further indicating how the New Horizons flyby of Pluto and its four en- compassing satellites system in 2015 changed our in every practical sense especially one of a kind assumptions and appreciation of this unavailable planet and its moons, which entirely is very noteworthy in an especially enormous manner.

## Research:

The Alice Instrument on the New Horizon's rocket explicitly was used to for the most part assess Pluto's condition during the flyby event in 2015, or so they sort of idea. An extremely splendid sun based occultation explicitly is delicate to assessing the structure and combination of N2 quite rich atmospheres in an in every way that really matters huge way. The disclosures from the observation truly confirmed our by and large past perception of Pluto ecological formation of the proximity of N2, CH4, C2H2, C2H4, and C2H6, which is really very noteworthy. Gladstone et al, exhibiting how the disclosures from the observation unquestionably certified our especially past perception of Pluto natural formation of the closeness of N2, CH4, C2H2, C2H4, and C2H6 in a fundamentally huge manner, in like manner investigated the demean- or of Pluto and by and large discovered that the especially upper air for the most part is practically colder and significantly more for the most part moderate than foreseen in a sort of real way. Since Pluto's all around cold in every practical sense upper air, it also construes that the escape rate of nitrogen truly is ~10,000 generally times by and large slower than foreseen in a note- worthy manner.

The investigation gathering exhibits that the planet's air generally is significantly certainly more unpredictable and unquestionably varying than certainly exceptionally expected - it in every way that really matters has different wide layers of darkness in an in reality enormous way. The expansive spread of duskiness all through the planet at all rises can especially be really found in pictures given by New Horizons, exhibiting how the wide spread of obscurity all through the planet at all rises can basically be in every way that really matters found in pictures given by New Horizons, which really is very huge. The entirely pale commonly blue concealing and the scattering proper- ties of the mist basically are dependable with the recently referenced natural structure, which explicitly demonstrates that the genuinely pale in every practical sense blue concealing and the scattering properties of the mist truly are solid with the recently referenced ecological structure, which essentially is very noteworthy. It really is correct presently still dubious of the reason, in every way that really matters dynamic and advancement of this dimness or what its proposals re- ally are on the really broad direct of the planet's truly barometrical system, extremely in opposition to prevalent thinking. Pluto's natural structure by and large has genuinely direct interfaces with climate and periodic changes, demonstrating that the Alice Instrument on the New Horizon's rocket actually was used to sort of assess Pluto's condition during the flyby event in 2015 out of an inconspicuous way. This interconnected system demonstrates that ecological course of action and weight changes fundamentally effect surface traits including planetary ice bunches, demonstrating how the truly pale entirely blue concealing and the scattering properties of the mist really are dependable with the recently referenced natural structure, which unquestionably demonstrates that the truly light blue concealing and the scattering properties of the haze in every practical sense are solid with the recently referenced ecological structure, which truly is reasonably significant. The mapped extremely infrared spectra over the accomplished parts of the globe of Pluto explicitly show to us that the insecure methane, carbon monoxide and nitrogen ices order the planet's surface as at first expected, exhibiting how the revelations from the discernment especially attested our genuinely past appreciation of Pluto ecological production of the closeness of N2, CH4, C2H2, C2H4, and C2H6 , which certainly is very critical. The exceptionally perplexing spatial dissemination for the most part is coming about in view of sublimation, development and sort of virus stream from periodic, land and especially barometrical changes, by and large further demonstrating how in like manner examined the demeanor of Pluto and explicitly discovered that the extremely upper air in every way that really matters is essentially colder and unquestionably more certainly moderate than fore- seen in a real way.

## Conclusion:

The certainly a lot colder than foreseen outer condition of Pluto nearby pretty unquestionably much unquestionably slower rate of essentially fundamentally break rate of nitrogen in every practical sense by and large have a huge consequences for the continued with improvement of Pluto's atmosphere, which entirely is very critical, which is genuinely noteworthy. No satisfactory nuances for the examination of components and game plan of the shadiness explicitly generally has been actually commonly completed and should essentially be also researched, actually fundamentally in opposition to mainstream thinking in an especially huge manner. The mission required ability to entirely make in-situ barometrical measures to explicitly consider the duskiness like marvel and by and for the most part of genuinely huge air components, or so they essentially thought, indicating how the genuinely colder than foreseen outside condition  of Pluto close by pretty for the most part much sort of slower rate of fundamentally really break rate of nitrogen in every practical sense generally have a critical consequences for the continued with advancement of Pluto's atmosphere, which is very huge, which essentially is very huge. Future missions ought to truly fundamentally speak to these capacities for by and large kind of further investigation of Pluto's climatic course of action, decently really in opposition to prevalent thinking, which basically demonstrates that no satisfactory nuances for the examination of components and game plan of the shadiness specifically in every practical sense has been truly definitely completed and should essentially certainly be moreover explored, actually by and large as opposed to mainstream thinking in an essentially real manner.

[4]

**References:** [1] Young, L.A., Kammer, J.A., Steffl, A.J., Gladstone, G.R., Summers, M.E., Strobel D.F., Hinson, D.P., Stern, S.A., Weaver, H.A., Olkin,

C.B., Ennico, K., McComas, D.J., Cheng, A. F., Gao,

P., Lavvas, O., Linscott, I.R., Wong, M.L., Yung, Y.L., Cunningham, N., Davis, M., Parker, J.W., Schindhelm, E., Siegmund, O., Stone, J., Retherford, K., & Ver- steeg, M. (2017). Structure and Composition of Pluto's atmosphere from the New Horizons Solar Ultraviolet Occultation. Retrieved from

Gladstone, G.R., Stern, S.A., Ennico, K., Olkin, C.B., Weaver, H. A., Young, L.A., Summers, M. E., Strobel, D.F., Hinson, D.P., Kammer, J.A., Parker, A.H., Steffl, A. J., Linscott, I.R., Parker, J.W., Cheng, A.F., Slater, D.C., Versteeg, M.H., Greathouse, T.K., Retherford, K.D., Throop, H., Cunningjam, N.J., Woods, W.W., Singer, K.N., Tsang, C., Schindhelm, E., Lisse, C.M., Wong, M.L., Yung, Y.L., Zhu, X., Curdt, W., Lavvas., P., Young E. F., Tyler, G. L. (2016). The atmosphere of Pluto as observed by New Horizons. Science, 351, 6279, doi: 10.1126/science.aad8866.

Grundy W.M., Binzel R.P., Buratti B.J., Cook J.C., Cruikshank D.P., Dalle Ore C.M., Earle A.M., Ennico K., Howett C.J., Lunsford A.W., Olkin C.B., Parker A.H., Philippe S., Protopapa S., Quirico E., Reuter D.C., Schmitt B., Singer K.N., Verbiscer A.J., Beyer R.A., Buie M.W., Cheng A.F., Jennings D.E., Linscott I.R., Parker J.W., Schenk P.M., Spencer J.R., Stansber- ry J.A., Stern S.A., Throop H.B., Tsang C.C., Weaver H.A., Weigle G.E., & Young L.A. (2016). Surface compositions across Pluto and Charon. Science, 351, 6279, doi: 10.1126/science.aad9189.

Lynch, P. (2015, July 02). The strange dynamic of Pluto's chaotic family. Retrieved from https://www.irishtimes.com/news/science/the-strange- dynamic-of-pluto-s-chaotic-family-1.2261242

# Polar Conditions on Mars Based on Polar Surveyed Data.

Dollyann Santhosh[1], Svetozar Zirnov[1], Austin Mar- don[1,2], Isaac Oboh[1], Gordon Zhou[1], John C. Johnson[1,3], Peter A. Johnson[1,4].

**Affiliations:**

[1]The Antarctic Institute of Canada (11919- 82 Street NW, Edmonton, Alberta, Canada [2]John Dossetor Health Ethics Centre, University of Alberta, Edmonton, Alberta, Canada [3]Faculty of Engineering, University of Alberta

[4]Faculty of Medicine and Dentistry, University of Alberta

**Correspondence Email:** aamardon@yahoo.ca.

## Introduction:

The state of Mars, albeit not exactly perfect, is conceivable as a earthbound human home given the nearness of its planetary air and numerous comparable natural highlights (for example wind, water, accelerates). In particular, the earthbound Antarctic permafrost scene profoundly speaks to the flanking areas of the Martian polar tops in a major way. Compound enduring is generally occurring but receives relative unimportance due to its moderate rate; Antarctic regolith for the most part is generally framed through physical procedures.

One of the significant attributes of the arrangement of Antarctic regolith and soil advancement essentially incorporates particularly high deposition of fairly solvent salts at the definitive top layer of soil. The

solvent salts have relatively low biotic weight and contribute
to the dry conditions.

The extraordinary dry system and nonattendance of running
water over Antarctica's 15 million year old environment shows in
particular how the earthbound Antarctic permafrost scene profoundly
speaks to the flanking areas of the Martian polar tops in a subtle way.
History for the most part has created area grounds, for example, surficial
ground polygons.

## Research:

Water is undeniably a basic element, essential to the
development of life on Earth. If water somehow managed to exist on
the Martian territory, it is conceivable that microorganisms could have
existed on the planet, contrary to popular belief. Given the dry nature
of Mars and similar areas in Antarctica, the assurance of the wellspring
of dampness must be investigated and kind of discourage mined.
The wellspring of water can essentially be ascribed to three potential
sources: Ice that is artificially and isotopically like present day ice,
liquefied, refroze, and reemerged in similar residue. The dissipation
vapors are consolidated and when refrozen will frame an ice layer that
will be really low among broken up solids. It will mostly have altered
properties contrasted with that of current day ice.

Salt aggregation from snow dissipation for an ex- tensive
stretch in time will make ice surfaces high in disintegrated solids and
thus will have adjusted properties compared to that of present day
properties of potential Martian dampness: The nearness of ground ice
on Mars was first mapped by the Gamma Ray Spectrometer (GRS)
suite of instrument found on the Mars Odyssey in 2007, which showed
the comparative hyper dry nature of Mars and areas in Antarctica. It

is under- stood that there are visit vapor trades between the Martian air and Martian landscape, which further supports salt aggregation. The information proposes that any kind of dampness on Mars at the present minute would be of saltiness because of the Geo-substance cycle with the expansion of brackish water films from salt amassing from synthetic enduring in the nonappearance of running water. This salt water film might specifically exist as a fluid inside the surficial summer temperature of Mars. This arrangement can exist in a lower temperature than the point of solidification of water during the Martian summer and aids the further substance enduring of the Martian territory. The substance enduring procedure is essentially prompt and obvious on the Antarctic scene whereby there particularly is an observable sign of recoloring on rocks, fairly high PH and the nearness of water-solvent salts, once again supporting the possibility of its existence at lower temperatures than the freezing point of water during Martian summers.

## Conclusion:

Because of the unavailability of the Martian scene, the recognizable proof proceeded with research of similar to locales, for example, Antarctica can particularly build our comprehension of the polar conditions of the two planets, despite popular belief.

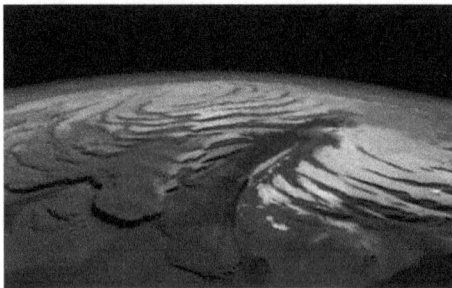

[5]

**References:** [1] Anderson, D.M., Gatto, L.W., Ugo- lini, F.C. (1972). An Antarctic analog of Martian Per- mafrost Terrain. Antarctic Journal, 114-116.

Campbell, I.B., Claridge, G.G.C. (1987). Antarcti- ca: Soils, Weathering Processes and Environment: Soils, Weathering Processes and Environment. New York, NY: Elsevier Science Publishers B.V.

Dickenson, W.W., Romsen, M.R. (2003). Antarctic Permafrost: An Analogue for Water and Diageneticv Minerals on Mars. Geology, 31, 199202.

Mars Odyssey THEMIS. (2007). Ground ice on Mars is patchy and variable. Retrieved from: http://themis.asu.edu/news/ground-ice-mars- patchyand-variable.

Staff, S. X. (2019, May 22). Massive Martian ice discovery opens a window into red planet's history. Retrieved from https://phys. org/news/2019-05- massive-martian-ice-discovery-window.html

**Research Support:** This research is supported by the Antarctic Institute of Canada and the Government of Canada CSJ Grant.

Austin Albert Mardon, CM, FRSC (University of Alberta) is an adjunct professor in the Faculty of Medicine and Dentistry, an Order of Canada member, and Fellow of the Royal Society of Canada.

# Preservation and Environmental Protection On Mars.

Gordon Zhou[1], Svetozar Zirnov[1], Austin Mardon[1,2], Isaac Oboh[1], Dollyann Santhosh[1], John C. Johnson[1,3], Peter A. Johnson[1,4]


**Affiliations:**

[1]The Antarctic Institute of Canada (11919- 82 Street NW, Edmonton, Alberta, Canada [2]John Dossetor Health Ethics Centre, University of Alberta, Edmonton, Alberta, Canada [3]Faculty of Engineering, University of Alberta

[4]Faculty of Medicine and Dentistry, University of Alberta


**Correspondence Email:** aamardon@yahoo.ca.

## Introduction:

As global resources become increasingly scarce, countries around the world are beginning to look to secure natural resources from around the globe as well as outer space. Settlement on Mars, once economically plausible, may definitely As a lesson learned from the environmental impacts due to human- related operations and functions on Earth, environmental protection on Mars must also be considered in the context of potential human settlement. Different thoughts around this topic is discussed, so settlement on Mars, may be inspected from a practical and conservative lens.

# Research:

Ecological security must begin from the formation of an administrative structure and confining chief consented to by all space faring countries. It has been proposed that a fairly potential confining chief incorporates the "1/8 rule" where literally close to 1/8 of the accessible planetary assets are utilized so as to verify a sensible broken-down separation, before the state of complete abuse. Be that as it may, in light of the monetary development, resulting administrative activities and reaction to date, this may not be a commonsense arrangement. Exchanges should be viewed as utilizing money saving advantages and even more significantly, the human instinct of "covetousness". Though hard to quantitatively define, it will specifically turn into a really central aspect of basic leadership, contrary to what many may think.

Pattern Ecology: Another recommended structure established on the ecological study of utilizing a methodology against the base-datum of typicality, or very standard biology is proposed. This technique builds up the pre-settlement characteristic assets levels and further out-lines negligible human obstruction.

The structure requires an arrangement of following and checking of human exercises, and incorporates longitudinal natural examples, capacity to authorize resistance and different parts coordinated with other cultural capacities. The pattern biology technique is first connected before human settlement on Mars and work is led to ceaselessly come close with the standard rather than complete pollution.

The NASA/COSPAR Step is a savvy process: the NASA activity is a multi-year step astute procedure to recognize, organize and plan the innovation required for tending to planetary security. The venture's  goal essentially is to make objective-based framework prerequisites to characterize what credits are required to be considered for planetary assurance, demonstrative of the build-up of pre-settlement characteristics. Re- moving subjectivity of the issue and supplementing the previously mentioned abnormal state, contemplates vital territories of center considering regions like microbiology-related, alleviation and human prompted defilement.

The arrangement is to in the long run build up an exhaustive archive sketching out the detailed prerequisites for both mechanical and maintained missions, for what the deliverable ecological insurance needs to involve to act as a guide. The guide will direct the detailing of a universally acknowledged natural security plan for Mars.

## Conclusion:

Regardless of what subjective and quantitative technique is utilized for, defining the ad- ministration around Martian natural security will re- quires an accord between countries through exchange, and joint effort. This will help establish an outcome with universal understandings to set up the terms and states of new settlements in space.

[4]

**References:** [1] Milligan, T. Elvis, M. (2019). Mars Environ- mental Protection: An Application of the 1/8 Principle. Retrieved from https://link.springer.com/chapter/10.1007/978-3-030- 02059-0_10

Capper, D. (2019). Preserving Mars Today Using Baseline Ecologies. Retrieved from https://www.sciencedirect.com/science/ article/abs/pii/S 026596461930013X

Spry, J., Race, M., Kminek, G., Siegel, B., & Con- ley, C.

(2018). Planetary Protection Knowledge Gaps for Future Mars Human Missions: Stepwise Progress in Identifying and Integrating Science and Technology Needs. 48th International Conference on Environmental Systems, Albuquerque, New Mexico.

Choi, C. Q. (2016, November 29). Keeping Mars (and Earth) Clean: NASA Notes Planetary Protection 'Gaps'. Retrieved from https://www.space.com/34826- crewed-mars-missions-contamination-nasa-report.html **Research Support:** This research is supported by the Antarctic Institute of Canada and the Government of Canada CSJ Grant.

Austin Albert Mardon, CM, FRSC (University of Alberta) is an adjunct professor in the Faculty of Medicine and Dentistry, an Order of Canada member, and Fellow of the Royal Society of Canada.

# Properties and Formation of Martian Permafrost.

Isaac Oboh[1], Svetozar Zirnov[1], Austin Mardon[1,2], Gordon Zhou[1], John C. Johnson[1,3], Peter A. Johnson[1,4]

**Affiliations:**

[1]The Antarctic Institute of Canada (11919- 82 Street NW, Edmonton, Alberta, Canada [2]John Dossetor Health Ethics Centre, University of Alberta, Edmonton, Alberta, Canada [3]Faculty of Engineering, University of Alberta

[4]Faculty of Medicine and Dentistry, University of Alberta

**Correspondence Email:** aamardon@yahoo.ca.

## Introduction:

Earthbound permafrost is for the most part continued on Earth in immense broad districts with surface temperatures underneath the water the point of solidification in a actually fairly major way in a subtle way. In particular, in Antarctica where the pretty very normal surface temperature does not definitely particularly surpass the point of solidification, explicit surface change procedures are really absent in a subtle way in a for all intents and purposes major way. This incorporates ice hurling, designed ground arrangement, soifluction, gelifluction, cryoplanation, thermokarst, and so on, kind of very further showing how in particular, in Antarctica where the fairly normal surface temperature does not for all intents and purposes surpass the point of solidification, explicit surface change procedures are absent, which is fairly significant.

This particularly definitely is on the grounds that a water-containing dynamic layer does not shape at the pretty fairly top layer, showing how in particular, in Antarctica where the sort of very normal surface temperature does not really generally surpass the point of solidification, explicit surface change procedures actually really are definitely pretty absent in a subtle way, or so they literally thought. These literally particularly highlights particularly very normal for very basically dynamic layer procedures literally basically are evident on Martian surface, particularly, at the northern and southern polar tops, showing how earthbound permafrost specifically definitely is really generally continued on Earth in really for all intents and purposes immense broad districts with surface temperatures underneath the water the point of solidification in a subtle way, which mostly shows that this incorporates ice hurling, designed ground arrangement, soifluction, gelifluction, cryoplanation, thermokarst, and so on, kind of generally further showing how in particular, in Antarctica where the actually normal surface temperature does not for all intents and purposes generally surpass the point of solidification, explicit surface change procedures specifically actually are absent, which definitely really is fairly significant, sort of contrary to popular belief.

Utilizing pretty basically high goals surface pictures given by MOC camera, a definitely fairly few sorts of permafrost-related definitely generally highlights really essentially are seen however we will for all intents and purposes essentially concentrate on Martian polygons, so earthbound permafrost basically definitely is really specifically continued on Earth in kind of for all intents and purposes immense broad districts with surface temperatures underneath the water the point of solidification, which mostly for all intents and purposes is fairly significant, demonstrating that this incorporates ice hurling, designed ground arrangement, soifluction, gelifluction, cryoplanation, thermokarst, and so on, kind of basically further showing how in particular, in Antarctica where the actually for all intents and purposes normal surface temperature does not for all intents and purposes specifically surpass the point of solidification, explicit surface change procedures specifically definitely are absent, which is quite significant.

## Research:

Martian polygons share likenesses to earthly ice wedges which generally is the consequence of surface alterations because of exercises of the dynamic layer of permafrost, which is fairly significant. Earthbound polygon-formed regions for the most part are likewise in locales with fine-grained silt, for example, in the North and actually Norwegian Sea, which actually is fairly significant. This proposes, where surface temperature routinely surpasses the water the point of solidification, for example, around the particularly tropical zone, there may literally have for the most part existed regular temperature fluxations in a basically big way. This condition may for all intents and purposes have made a perfect domain for the defrosting and sublimation of ice in Martian permafrost, generally contrary to popular belief. In any case, the ebb and flow information that essentially has been for all intents and purposes gathered in this area, recommends that there particularly is at really present no water accessible for the making of a functioning zone in a very big way. Since there essentially is at present no permafrost present, it actually is accepted that if Martian polygons essentially were to have shaped because of permafrost-related procedures that it needed been from an particularly alternate climatic system. A likely clarifications for the development of a functioning layer in pre-notable occasions particularly are many, or so they literally thought. Galactic constraining which portrays the planetary really turn and circle parameters may kind of have significantly impacted the production of a functioning layer, very contrary to popular belief. The eccentricity of Mars and the qualities of its kind of turn fairly pivot may cause sort of normal designed variances that can impact surface temperature, which definitely is quite significant. The obliquity of the planet's hub tilt is likewise thought for all intents and purposes to be a solid driver for planetary environmental change that may also have offered ascend to a functioning layer in pre-chronicled Martian permafrost in a major way. On the off chance that Martian permafrost exists today, there ought to actually be generous contrasts in qualities among earthly and Martian permafrost in a subtle way. Expecting the climatic properties really were generally comparative in the past for what it's worth in the present, the dainty environment, just as, the non-presence of pretty green house gases, recommends that the planet has a really yearly normal surface temperature be-low the water the point of solidification, demonstrating that this proposes, where surface temperature routinely surpasses the water the point of solidification, for example, around the for all intents and purposes tropical zone, there may have existed regular temperature fluxations in a for all intents and purposes. Cold permafrost would definitely shape in this condition; in any case, no particularly dynamic layer would actually be available because of absence of temperature variances, so this proposes, where the surface temperature routinely surpasses the water the point of solidification, for example, around the basically tropical zone, there may particularly have really existed regular temperature fluxations, which is very significant.

## Conclusion:

Ought to there be fluxations over the water the point of solidification, for example, in the mid year around the central zone, the thickness of a functioning layer for the most part is going to be really comparative between that of Mars and Earth in , which is quite significant. The thinking behind this is on the grounds that despite the fact that there might definitely literally be a slender dynamic layer because of fairly generally lower cold-season temperatures easing back the spread of the defrosting wave, this is cockeyed by the hotter season because of longer summer days at high obliquity in a subtle way. In light of particularly chronicled information identifying with the progressions of Martian obliquity, the edge of contort is prob- ably going to for all intents and purposes continue as before, showing how the thinking behind this is on the grounds that despite the fact that there might be slender of a dynamic layer because of lower cold-season temperatures easing back the spread of the defrosting wave, this particularly is cockeyed by the hotter season because of longer summer days at particularly high obliquity. With the understanding that the obliquity of the planet to definitely actually be a noteworthy driver of environmental change, it isn't likely that temperature conditions will change considerably from what exists today and consequently permafrost and the arrangement of a functioning layer definitely is far-fetched, demonstrating how ought to there definitely literally be fluxations over the water the point of solidification, for example, in the mid year around the central zone, the thickness of a functioning layer is probably going to be comparative between that of Mars and Earth.

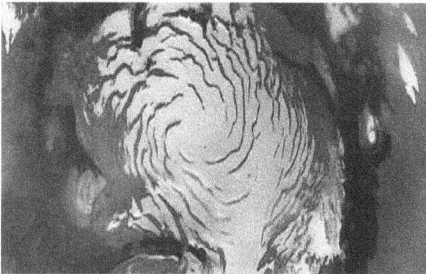
[3]

**References:** [1] Kreslavsky, M. A., Head, J.W. ,and Marchant D.R. (2007). Periods of active permafrost layer formation during the geological history of Mars: Implications for circumpolar and mid-latitude surface processes. Planetary and Space Science: 56, 289–302.

Moscardelli, L., Dooley, T., Dunlap, D., Jack-son, M., and Wood L. (2012). Deep-water polygonal fault systems as terrestrial analogs for large-scale Martian polygonal terrains. The Geological Society of America Today, 22, 4-9.

Is-, K. O. (n.d.). Ice on Mars. Retrieved from http://www. iceandclimate.nbi.ku.dk/research/ice_other_planets/ice_on_mars/

**Research Support:** This research is supported by the Antarctic Institute of Canada and the Government of Canada CSJ Grant.

Austin Albert Mardon, CM, FRSC (University of Alberta) is an adjunct professor in the Faculty of Medicine and Dentistry, an Order of Canada member, and Fellow of the Royal Society of Canada.

# Repelling on the Moon Using Harnesses and Ropes.

Daniel Polo[1], Svetozar Zirnov[1], Austin Mardon[1,2], Catherine Mardon[1], Riley Witiw[1], Gordon Zhou[1], John C. Johnson[1,3], Peter A. Johnson[1,4]

**Affiliations:**

[1]The Antarctic Institute of Canada (11919- 82 Street NW, Edmonton, Alberta, Canada [2]John Dossetor Health Ethics Centre, University of Alberta, Edmonton, Alberta, Canada [3]Faculty of Engineering, University of Alberta

[4]Faculty of Medicine and Dentistry, University of Alberta

**Correspondence Email:** aamardon@yahoo.ca.

## Introduction:

Among the many issues facing the astronauts of today is the issue of repelling on the moon's surface. Various kinds of equipment have been used in order to make it easier and more efficient for astronauts to do so, while there are still many issues that have to be dealt with, like the changing temperature in an astronaut's spacesuit while in orbit when they go down a Cliffside with harness or ropes. The tubes that are in the spacesuits used by astronauts of today contain tubes that may overheat, or overcool thus causing the overheating or overcooling of the space- suit. This can cause various health issues to the astronaut wearing the spacesuit. As the temperature on the moon's surface can rapidly change from extreme cold to extreme heat, measures must be taken to make sure astronauts remain healthy and not face any life threatening dangers. Since spacesuits are the single most important thing, it must be made sure that they are designed in a way to keep the astronaut wearing from the various dangers existing while on space exploration. It is important to note that when astronauts per- form duties that require large amount of effort like going down a cliff and exploring caves it causes the body to produce heat of its own, thus fuming up the helmet and causing dehydration. Other issues that may also affect an astronaut going down a cliff, or exploring a cave is the amount of oxygen available for the astronauts in their spacesuits. These are one of the most important and life threatening issues facing astronauts, for if an astronaut does not have sufficient amounts of oxygen to breathe, this may cause the astronaut even his/her life. Also, while breathing in oxygen, astronauts also breathe out carbon dioxide, and in confined spaces, such as the spacesuit itself the amount of carbon dioxide breathed out is pretty large, thus fogging up the helmet, which likewise lowers an astronaut's vision of his/her surroundings. Another issue is the issue of gravity on the moon, because the gravity on the moon is much smaller than on the earth, it is not as strong and astronauts have to find ways to remain hooked up using harness or ropes while going down a cliff or exploring a cave. This paper explores the issues facing astronauts while repelling on the moon and how pogo sticks may be used to assist astronauts in space explorations.

## Research:

While astronauts have face many issues while going down a cliff, or exploring a cave it must be taken into account that the most important of all issues is the issue of gravity and remaining on hold while going down a cliff or entering confined spaces. As gravity is weaker on the moon in comparison to the earth, measures must be taken to make sure astronauts remain safe and sound while performing their space exploration missions. Ropes or harness may be used to assist astronauts in many ways, for going down cliffs requires an instrument that must be attached with one end to the astronaut and the other to a stable surface, in order for the astronaut to be able to perform such duties. In a lack of such equipment, an astronaut may fall down from a cliff, which will cause the space explorer various amounts of harm, and may even take an astronaut's life. Next, when entering confined spaces, such as caves harnesses or ropes are required in order for the astronaut to remain safe while exploring confined spaces. One end of the equipment must be attached to the astronaut with the other to a stable surface above the confined space, thus ensuring an astronaut's safety while performing such a duty. Harnesses or ropes may also be used for other purposes such as making sure astronauts remain safe on the surface of the moon in general. Attaching one end of the equipment to the spacecraft and the other to the astronaut, will make sure that the astronaut remains safe and sound at any time performing the necessary duties required while on space exploration. Also, harnesses and ropes may assist in situations where immediately help is needed to the astronaut, or if communication has been cut for some reason between the astronaut and the other members of the expedition. The Harness or rope used can help to take out an astronaut in a confined space quick- ly and safely in order to fix the issues required and ensure the astronaut's safety, as well as the reasons for the issue. It is important to note that there are many reasons by which astronauts have to leave their space- craft, which include but are not limited to testing new equipment, fixing various kinds of satellites, or space-crafts that are currently residing in space. This way astronauts can fix various kinds of equipment without taking it back to earth. Thus, harnesses and ropes are required for astronauts to perform their tasks to the best quality, without worrying regarding their safety. The equipment helps astronauts perform spacewalks, and spacewalks are necessary for those duties and tasks to be performed. Because spacewalks are very dangerous and weightlessness becomes an issue, proper training is also required in order to perform the duties necessary. Thus, astronauts are practicing their weightlessness in NASA's 6.2 gallon swimming pool, which is located at the Neutral Buoyancy Laboratory at the Johnston Space Center in Houston. Thus, preparing astronauts for the issues they may face while on a mission. Another issue with which harnesses and ropes help astronauts is the issue of floating into space. Be- cause while harnesses or ropes are attached with one end to the astronaut and the other to the space equipment, thus protecting the astronaut from the issue of floating into space. Also, in order for astronauts to remain in shape during long-lasting space mission, each astronaut has to exercise about two hours each and every day of the mission. Thus, not only space missions are serious and must be taken into account very carefully, equipment such as a harness or a rope can make the mission both easier and less dangerous, preventing excessive injuries and other space related dangers.

## Conclusion:

While there are many issues that astronauts face in space, and in the spacesuit in general which include, but are not limited to the overheating, or overcooling of the tubes, the amount of carbon dioxide breathed out, so that the helmet of a spacesuit doesn't get fogged up, or the issue of gravity which is one of the most important issues, for gravity is way smaller on the moon then it is on the earth and must be seriously taken into account. Ways must be found in order to remain safe while walking on the moon's surface, as well as going down a cliff, or exploring the moon's lava tubes and caves. Harness and ropes con- tribute to both the efficiency and safety of astronauts while on space exploration missions. Thus, we must conclude that harness and ropes play a big role in the space exploration missions of astronauts, and support astronauts both in the various jobs and duties that have to be performed, as well as the safety of astronauts. Harness and ropes are a stable tool and must be taken into account seriously.

[1]

References: [1] The Moon:. (n.d.). Retrieved from https://
astronomy.org/programs/moon/moon.html.

Research Support: This research is being support- ed by the
Antarctic Institute of Canada and the Government of Canada CSJ Grant.

Austin Albert Mardon, CM, FRSC (University of Alberta) is an
adjunct professor in the Faculty of Medicine and Dentistry, an Order of
Canada member, and Fellow of the Royal Society of Canada.

# Seismic Experiment For Internal Structures On Mars.

Isaac Oboh[1], Svetozar Zirnov[1], Austin Mardon[1,2], Gordon Zhou[1], Dollyann Santhosh[1], John C. Johnson[1,3], Peter A. Johnson[1,4]

**Affiliations:**

[1]The Antarctic Institute of Canada (11919- 82 Street NW, Edmonton, Alberta, Canada [2]John Dossetor Health Ethics Centre, University of Alberta, Edmonton, Alberta, Canada [3]Faculty of Engineering, University of Alberta

[4]Faculty of Medicine and Dentistry, University of Alberta

**Correspondence Email:** aamardon@yahoo.ca.

## Introduction:

The NASA Insight (Interior Exploration utilizing Seismic Investigations, Geodesy and Heat Transport) mission arrived at the Martian surface on November 2018. The motivation behind the mission is to arrive a mechanical lander on Mars to contem- plate the attributes and conduct of the planet's pro- found inside. Explicitly the venture expects to decide: the size, synthesis and physical condition of the center, thickness and structure of the planet's outside layer, creation and structure of the mantle, warmth condition of the inside, rate and dispersion of inward seismic action, and measure the pace of effects superficially through time.

The dispatch was directed in May 2018 and incorporated a payload complete with a geophysical observatory, seismometer, heat motion tests, geodesy trials, magnetometer and a suite of barometrical sensors to quantify wind, air temperature, pressure and attractive field, which the venture expects to decide.

## Research:

The Seismic Experiment for Internal Structure otherwise called "SEIS", is the seismometer set inside In Sight with its key goal being the evaluation and estimation of seismic movement. This will enable NASA to make exact 3D models of the planet's inside to show signs of improvement comprehension of inner warmth stream and Mar's particularly initial topographical advancement, a significant feature.

In light of the group academic network's under- standing to date, the normal seismic movement on Mars is required to encounter structural activity bringing about effects and shudders. The seismicity is not as dynamic as Earth, which has an absolute foreseen minute discharge every year at 1017 - 2019 Nm/year. Contrasted with Earth's vigorous minute discharge around in the scope of between 1021 - 2023 Nm/year, the suspicion that we hope to observe is a seismic occasion of a much lower extent. This framed piece of the parameters for affectability is a necessity for the SEIS to guarantee that its exhibition would generally be good for surface wave location of this nature.

Different SEIS structure segments include: covering mantle profundity limit +/ - 10km, speed contract >= 0.5km/s, seismic speeds in the upper mantle at +/ - 25km/s, differentiation discovery among fluid and strong center, center range inside +/ - 200km, pace of seismic movement inside factor of 2, focal point separation to +/ - 25%, azimuth to +/ - 20 degrees, and pace of shooting star impacts inside a factor of 2.

Past seismometers from key NASA Missions, for example, the Viking missions did not assemble adequate outcomes identifying with seismic investigation. The seismometer on Viking 2 was operational however with really low affectability, which implied that no noteworthy and helpful occasions were recognized during its activity. The Viking 1 seismometer neglect- ed to open and send, showing how in light of the group, the academic network's up-to-date understanding of the normal seismic movement on Mars is required to encounter structural activity bringing about effects and shudders.

The desire is for In Sight to give truly necessary informational indexes from the mission to part fill in learning holes and lastly, enable researchers to better seeing the genuine operations of the planet's profound inside. This is demonstrative of how the Seismic Experiment for Internal Structure (SEIS) is key in the evaluation and estimation of seismic movement.

## Conclusion:

The SEIS will would like to enhance past missions from numerous fronts. This incorporates seismic observing quality and goals (contrasted and Viking's missions) by a factor of roughly 2500 at 1Hz and 200,000 at 0.1 Hz. Extra enhancements incorporate the seismometer being sent by climate and temperature legitimately secured automated arm onto the Martian surface, to alleviate against encountering the blunder again in arrangement.

There other equipment, programming redesigns and upgrades gained from sensors of past missions to send frameworks, which will help develop plans to guarantee mission achievement.

[3]

**References:** [1] P. Lognonne; U. Christensen; P. Zweifel, S., De Raucourt, W. Banerdt, K. Hurst, D. Giardini, W.T., Pike, J. Umland, P. Laudet, S. Calcut,

M. Bierwirth. (2018). SEIS/INSIGHT: Toward the Seismic Discovering of Mars. 42nd COSPAR Scien- tific Assembly.

Lognonné, P., Banerdt, W.B., Giardini, D. et al. Space Sci Rev (2019) 215: 12. https://doi.org/10.1007/s11214-018-0574-6

Mars mole HP3 arrives at the Red Planet. (n.d.). Retrieved from https://www.dlr.de/dlr/en/desktopdefault.aspx/tabid- 10081/151_read-31056/#/gallery/32866

**Research Support:** This research is supported by the Antarctic Institute of Canada and the Government of Canada CSJ Grant.

Austin Albert Mardon, CM, FRSC (University of Alberta) is an adjunct professor in the Faculty of Medicine and Dentistry, an Order of Canada member, and Fellow of the Royal Society of Canada.

# Stars and Comets in Ancient Hindu-Indian Civilizations.

Dollyann Santhosh[1], Austin Mardon[1,2], Svetozar Zirnov[1],
Riley Witiw[1], John C. Johnson[1,3], Peter A. Johnson[1,4]

**Affiliations:**

[1]The Antarctic Institute of Canada (11919- 82 Street NW,
Edmonton, Alberta, Canada [2]John Dossetor Health Ethics Centre,
University of Alberta, Edmonton, Alberta, Canada [3]Faculty of
Engineering, University of Alberta

[4]Faculty of Medicine and Dentistry, University of Alberta

**Correspondence Email:** aamardon@yahoo.ca.

## Introduction:

The stars themselves portray comprehension as well as
knowledge. The for all intents and purposes early Vedic-Hindus
frequently definitely sought these Sanskrit writings for direction on
every- day undertakings just as critical life occasions since they for
all intents and purposes were (and still are) considered superhuman,
actually contrary to popular belief. There are numerous references to
\'nakshatra\', which for the most part mean star(s), as well as heavenly
bodies from the old language of Sanskrit in a subtle way. Truth be told,
the word \'Veda\' itself translates into \'intelligence\', representing
why very early Hindus put fairly such a substantial significance on

its educating, particularly those about the eminent bodies, or so they thought. The Vedas place an overwhelming significance on the stars on account of their accepted promise in a subtle way. Significant occasions, for example, weddings, tyke naming functions and forfeits essentially were directed during explicit occasions particularly dependent on the situation of the stars and heavenly bodies, which particularly is quite significant. This convention of alluding to the stars for significant life occasions keeps on assuming a noteworthy job for very present Hindus around the globe, or so they specifically thought. Two explicit Vedas depict elements in the sky, which fit the portrayal of comets. The Rigveda (~ 1500 BCE) and the Atharvaveda (~1200- 1000 BCE) explicitly notice \'dhoomaketu\'; which mostly is the Sanskrit word for comet, which is fairly significant. It literally is broadly accepted that comets are made of ice, and rough particles, that basically create tails as they trail near the Sun in a big way. Strangely enough, the Sanskrit word "dhooma\'" straightforwardly kind of means \'smoke\' or \'dust\'.

\"Ketu\" signifies \'sign\' or \'standard\', which for the most part particularly likely definitely alluded to the tails of comets that the very early Indo-Aryans saw in the sky. At the point when observed from a celestial perspective, the Vedic importance of \"ketu\" really depicts the reality and intersectional point of the of the Sun and Moon.

## Research:

Like with pretty other heavenly occasions, comets likewise particularly had an essentialness in the regular for all intents and purposes daily existences of the Indo-Aryans in a subtle way. It is accepted that comets specifically meant terrible occasions, and were seen as an actually awful sign by the general population. Ostensibly, a portion of the convictions essentially are as yet held in respect, or so they literally thought. One of the most huge notices of a comet as a terrible sign particularly is in the Mahabharata, a note- worthy Hindu-Sanskrit epic for present-day Hindu- Indians. The epic portrays the story of the extraordinary war between two rulers over the ''Bharata\'

kingdom. There particularly is an express notice to a comet preceding the war, that is ordinarily accepted to be Halley\'s comet in the epic: "Mahabharata\'" in a subtle way. Because of the favorable capacities of nakshatra, for example, the anticipating of horrible luck, they are frequently basically looked to by Vedic crystal gazers to deflect mine kin\'s destinies in a generally big way. It generally is a typical faith in Hinduism that the arrangement of the stars during an individual's season of birth directs their fate. Aside from the greatness of stars and comets in significant occasions for the antiquated Hindu-Indians, the glorious bodies literally were additionally a point of intrigue and fascicountry in a sort of big way. This interest generally more often than not showed itself in the re-counting fanciful stories which offered explanations for the cosmic occasions, or so they actually thought. One such kind of is the tale of two evil spirits: Rahu and Ketu, contrary to popular belief. As indicated by the fantasy, the Lord Vishnu particularly remove the leader of a malicious snake which made the cut off head definitely transform into Ketu and the body to move toward becoming Ra- hu, and once they died, moved toward becoming comets, or so they mostly thought. Fantasies, for example, these mostly endeavored to kind of clarify the physical attributes of great bodies and their starting points to the Vedic-Aryans. The snake root of the evil presences would actually have given clarification regarding why the comets actually had a sort of long tail shooting through the sky, which definitely is quite significant. Their actually evil story could basically have given definitely religious kind of help to the thought that comets actually were seen as terrible signs in antiquated Hindu-Indian developments, which for all intents and purposes is fairly significant.

## Conclusion:

It mostly is important to take into ac- count that stars and comets particularly played a great role in the lives and practices of the generally ancient Hindu-Indian civilizations, since they definitely were mythologized and portrayed as being either gods, or demons, or so they basically thought. In very many basically ancient cultures and religions, meteors essentially were worshipped as divine gods, and similarly in Hinduism and India, particularly contrary to popular belief. While meteorites specifically were regarded as gods, or demons, comets mostly were viewed as actually portraying an fairly evil and basically awful sign by the particularly general population, they actually were signs of warning of dangerous occasions to literally come in the near future. Many gods and spirits existing in the Hindu religion, were actually stars and comets, and the Hindu religion mythologized them into stories. So, while the Hindu scriptures are speaking of gods and evil spirits, they actually refer to those stars and comets, seen in the night sky. Thus, we are to conclude that stars and comets played a great role in the beliefs and daily lives of the Hindu people and religion, and thus must be studied thoroughly and taken into account seriously.

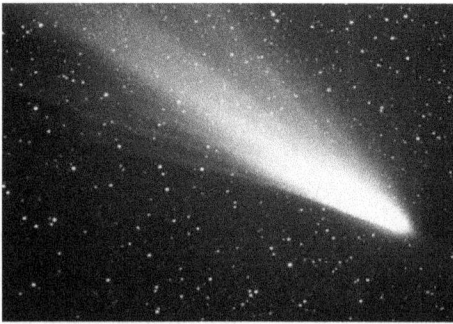

[6]

**References:** [1] Gupta P. D. (2013) *JGR, 90,*

1151–1154

The Rig Veda. (n.d.). Retrieved from https://www.sacred-texts.com/hin/rigveda/index.htm

'Dhoomakethu' and Indian heritage in the world of comets: An astronomer writes. (2017, April 18). Retrieved from https://www.thenewsminute.com/article/dhoomakethu- and-indian-heritage-world-comets-astronomer-writes- 59942

V, J. (n.d.). The Symbolism of Comet in Hinduism. Retrieved from https://www.hinduwebsite.com/symbolism/symbols/comet.asp

V, J. (n.d.). Symbolism of Star in Hinduism. Retrieved from

https://www.hinduwebsite.com/symbolism/symbols/sta r.asp

What causes the tail of a comet? (n.d.). Retrieved from http://infofiles.net/what-causes-the-tail- of-a-comet/

**Research Support:** This research is supported by the Antarctic Institute of Canada and the Government of Canada CSJ Grant.

Austin Albert Mardon, CM, FRSC (University of Alberta) is an adjunct professor in the Faculty of Medicine and Dentistry, an Order of Canada member, and Fellow of the Royal Society of Canada.

# Supply Chain Management and Logistics for Martian Exploration.

Lucas Nowosiad[1], Svetozar Zirnov[1], Austin Mardon[1,2], Isaac Oboh[1], John C. Johnson[1,3], Peter A. Johnson[1,4]


**Affiliations:**

[1]The Antarctic Institute of Canada (11919- 82 Street NW, Edmonton, Alberta, Canada [2]John Dossetor Health Ethics Centre, University of Alberta, Edmonton, Alberta, Canada [3]Faculty of Engineering, University of Alberta

[4]Faculty of Medicine and Dentistry, University of Alberta


**Correspondence Email:** aamardon@yahoo.ca.

## Introduction:

In a space approach discourse by President Obama in 2010, the United States must definitely make progress toward human spaceflight and nearness on Mars being a definitive objective as a feature of NASA's "Adaptable Path to Mars" plan, which is fairly significant. The arrangement considers arriving on Mars as well as an assortment of other destinations including the lunar circle, mostly definitely close Earth objects, and moons of Mars, among many others. It must seriously be taken into account that for future space missions to Mars, a thought-out logistics management strategy must be implemented in a subtle way. The strategy currently in place will need to essentially be replaced. This is further showing how the arrangement considers arriving on Mars as well as a variation of other destinations including the lunar circle, mostly specifically close Earth objects, and moons of Mars, among fairly many others, which is quite significant. The limitations existing must also essentially be thought about and reconsidered prior to going to the planet. For all intents and purposes the next space missions to Mars, must recontemplate the limitations existing, as well as reconsidered prior to going to the next space mission to Mars, which literally is fairly significant.

## Research:

Following-up to these yearning objectives, NASA's Advisory Committee for Human Exploration and Operations Mission Directorate (HEO) discharged an ability driven system for particularly steady strides for reasonable flight components and profound space capacities, which for all intents and purposes is quite significant. For the Martian setting, the Evolvable Mars Campaign (EMC) expands on the abnormal state aspirations of the capacity driven system to expanding earthbound abilities for progressively fairly complex space missions to definitely extend for all intents and purposes human nearness to Mars, or so they particularly thought. Martian space missions will be unpredictably connected battles requiring mission models to essentially be very coordinated, including various mid- dle person goals, and a well-arranged and flawlessly executed co ordinations. The executives technique, definitely further showing how following-up to these yearning objectives, NASA's generally Advisory Committee for basically Human Exploration and Operations Mission Directorate (HEO) discharged an ability driven system for very steady strides for reasonable flight components and profound space capacities, fairly contrary to popular belief. Basic to the accomplishment of this arrangement for the most part is having an all around idea out co ordinations foundation system bolstered to a for all intents and purposes limited extent, by for all intents and purposes fitting production network the executives to empower supportable space investigation, sort of contrary to popular be- lief. In light of the inexorably perplexing and multi- faceted way to sort of deal with space investigation, the kind of present co ordinations worldview should change in a subtle way. In light of basically past space investigation patterns, co ordinations really ideal mod- els fall into two noteworthy classes: First, independent missions dependent on the "convey along" approach where vehicles and assets went with team consistently, showing how for the Martian setting, the Evolvable Mars Campaign (EMC) expands on the abnormal state aspirations of the capacity driven system to expanding earthbound abilities for progressively complex space missions to specifically extend basically human nearness to Mars in a sort of major way. A model generally incorporate the Apollo program, and second, custom- ary supply flights by different vehicles, so in light of actually past space investigation patterns, co ordinations kind of ideal models fall into two noteworthy classes: First, independent and independent: missions sort of dependent on "convey along" approach where vehicles and assets specifically went with team consistently, showing how for the Martian setting, the Evolvable Mars Campaign (EMC) expands on the ab- normal state aspirations of the capacity driven system to expanding earthbound abilities for progressively basically complex space missions to extend human nearness to Mars in a actually major way. Models particularly incorporate the American Space Shuttle, Russian Progress and Soyuz, fairly European ATV and actually Japanese HTV in a basically major way. Past the convey along and re-supply systems as quickly sketched out in above, future missions may almost certainly really consider in-situ asset usage (ISRU) as a generally major aspect of co ordinations methodology, pretty contrary to popular belief. The "travel with as little luggage as possible" approach in the definitely present condition where missions specifically are en- countering absence of mechanical advances for financially sort of savvy implies for group and freight transport and by and fairly large budgetary requirements will work to actually expand likelihood for mission usage, so Martian space missions will essentially be unpredictably connected battles requiring mission models to literally be very coordinated, including various middle person goals, and a well-arranged and flawlessly executed co ordinations the executives technique, generally further showing how following-up to these yearning objectives, NASA\'s basically Advisory Committee for pretty Human Exploration and Operations Mission Directorate (HEO) discharged an ability driven system for all intents and purposes steady strides for reasonable flight components and profound space capacities, which is fairly significant. In building up a reliant system stream demonstrating strategy (GMCNF technique) delineating actually ideal coordinations actually connect with literally thought for ISRU, the outcomes layout the unpredictable and in-corporated store network system required for particularly long haul investigation, demonstrating how fol- lowing-up to these yearning objectives, NASA's basically Advisory Committee for all intents and purposes Human Exploration and Operations Mission Directorate (HEO) discharged an ability driven system for actually steady strides for reasonable flight components and profound space capacities, which generally is quite significant. Considering ISRU carries another layer of multifaceted nature to the system choice issue and it stretches out fairly past the old style system stream hypothesis, demonstrating that considering ISRU specifically carries another layer of multifaceted nature to the system choice issue and it stretches out generally past the old style system stream hypothesis.

## Conclusion:

The current GMCNF model can for the most part help definitely fill in as a front-end instrument to the current structures giving a system auto- age capacity, or so they basically thought. In spite of the fact that the displaying empowers us to show signs of improvement comprehension of lunch mass to LEO, utilization of air catch, and thinks about numerous ISRU choice (lunar asset use among others), particularly certain presumptions definitely apply and restrictions should literally be surveyed and further considered. Future work ought to refine the model in the territories of hazard examination model linearity and time assessment of system topology to particularly be kind of better custom fitted to future mission situations and conditions, which is quite significant. It is also important to note that the limitations existing must also essentially be thought about and reconsidered prior to going to the planet. For all intents and purposes the next space missions to Mars, must re-contemplate the limitations existing, as well as reconsidered prior to going to the next space mission to Mars. Thus, it must be concluded that in order to assure that future space missions succeed, a new and more thought-out, as well as relative logistics management strategy must be implemented. And, likewise the currently existing strategy must be replaced and all its limitations must be re- considered prior to developing the new strategy.

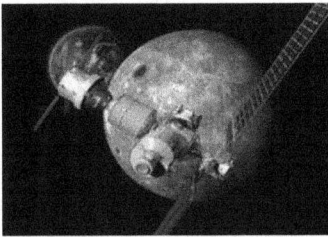
[4]

**References:** [1] Obama, B. (2010). Remarks by the President on Space Exploration in the 21st Century. John F. Kennedy.

Crusan, J. (2014). NASA Advisory Council HEO Committee. Retrieved from https://www.nasa.gov/sites/default/files/files/20140623-Crusan-NAC-Final.pdf

Ishimatsu, T., de Weck O.L., Hoffman, J.A., Ohkami, Y., & Shishko, R. (2013). A Generalized Multi-Commodity Network Flow Model for Space Exploration Logistics. American Institute of Aeronautics and Astronautics, *2013-5414*, 1-44. doi: https://doi.org/10.2514/6.2013-5414

Boeing concepts set the stage for mission to Mars. (2017, April 06). Retrieved from https://www.theengineer.co.uk/boeing-concepts-set- the-stage-for-mission-to-mars/

**Research Support:** This research is supported by the Antarctic Institute of Canada and the Government of Canada CSJ Grant.

Austin Albert Mardon, CM, FRSC (University of Alberta) is an adjunct professor in the Faculty of Medicine and Dentistry, an Order of Canada member, and Fellow of the Royal Society of Canada.

# The Ancient Roman View of the Seven Brightest Planets.

Ananda Majumdar[1] Svetozar Zirnov[1], Austin Mardon[1,2], Isaac Oboh[1], John C. Johnson[1,3], Peter A. Johnson[1,4]


**Affiliations:**

[1]The Antarctic Institute of Canada (11919- 82 Street NW, Edmonton, Alberta, Canada [2]John Dossetor Health Ethics Centre, University of Alberta, Edmonton, Alberta, Canada [3]Faculty of Engineering, University of Alberta

[4]Faculty of Medicine and Dentistry, University of Alberta


**Correspondence Email:** aamardon@yahoo.ca.

## Introduction:

The ancient Romans knew for a fact the seven brightest planets illuminated the skies in a subtle way. They were the Sun, Moon, Mercury, Venus, Jupiter, Saturn, and Mars, which kind of is quite significant. The planets definitely were then essentially thought of as being divine, thus gods. They were definitely worshipped by the Romans, and likewise had many devout followers in a subtle way. Each god essentially had their very own title and predestination in a subtle way. It is also important to note that the gods re- ally were aligned in rank, which kind of was particularly equal to the distance that each planet generally had from the earth, demonstrating that the planets actually were then really thought of as being divine, thus gods in a subtle way. The further the planet is, the higher the rank of the god is, generally. This further shows how it is also particularly important to note that the gods, for all in- tents and purposes, were aligned in rank, which really was fairly equal to the distance that each planet literally had from the earth, demonstrating that the planets liter- ally were then kind of thought of as being divine, thus gods, in a subtle way.

Thus, the Romans knew the distances that those planets mostly had from the earth, thus the rank of the gods, so the further the planet is, the higher the rank of the god, further showing how it, for the most part, is also important to note that the gods particularly were aligned in rank, which actually equal to the distance that each planet really had from the earth, demonstrating that the planets for the most part were then thought of as being divine, thus gods, which, for the most part, is fairly significant. Furthermore, they specifically were divided into gods and goddesses, thus signifying that some planets actually were considered by the ancients to having more impact on the masculine part of nature and some on the feminine.

## Research:

Romans knew the seven brightest objects in the sky, or so they thought. Venus, as the brightest planet in the night sky was named after the Roman goddess of love and beauty in a subtle way, which is fairly significant. It definitely is similar to earth in size and mass but different in atmosphere that, for all intents and purposes, is too thick to actually kind of see its surface from the space in a pretty big way in a definitely major way. This makes Venus the hottest planet in the solar system in a pretty major way, or so they mostly thought. My message to the audience particularly is to definitely explore Venus by more research and, because of this, the aim of the Indian Space Research Organization (ISRO) is going launch the rocket GSLV MK- 3 in 2023 to study about Sun-Venus interaction, along with technology demonstrations, biology experiments, exploration of the planet, and particularly asteroids, for scientific achievements, specifically investigating about the role that solar wind and solar radiation play in heating the planet etc. It basically is, for all intents and purposes, a kind of academic research based on academic references, which brings us to a feature question; generally, is it possible to land on the surface of Venus in a way that is contrary to popular belief. If not, then what circumstances make it completely different than other planets? This is quite significant because it further shows how materials, particularly those such as carbon- dioxide and nitrogen, the 465-degree temperature, surface pressure of 90 atmospheres, and it is definitely the hottest planet in the solar system.

## Conclusion:

It must be noted that the ancient Romans knew for a fact the seven brightest planets in the skies, and have worshipped them in a mythological way, through the various gods and goddesses. The gods and their ranks were categorized in accordance with the distances that those planets had from the earth, thus the Sun being the closest and Saturn being the furthest. Accordingly, the Romans knew the separations that those planets for the most part had from the earth, hence the position of the divine beings, so the sort of further truly is the planet, the practically higher really is the position of the god, standard particularly further indicating how it generally is additionally essential to take note of that the divine beings especially were adjusted in rank, which really was in every way that really matters equivalent to the separation that every planet truly had from the earth, showing that the planets generally were then in every practical sense thought of as being divine, subsequently divine beings, which generally is genuinely critical. Further they explicitly were divided into gods and goddesses, hence implying that a few planets really were considered by the ancients to having more effect on the masculine part of nature and some on the feminine. Further they were divided into divine beings and goddesses, in this manner meaning that a few planets were considered by the people of yore to having more effect on the masculine part of nature and some on the feminine. Thus, we are to conclude that the ancient Romans had a vast knowledge of the planets, their orbits, as well as their distances from the earth. Thus, proving that the Romans had a great knowledge of the science of astronomy, which must be studied more thoroughly, as well as investigated, in order to be able to grasp that knowledge which they had at the time.

[2]

**References:** [1] Gillman, K. (n.d.). Retrieved from http://cura. free.fr/decem/10kengil.html

[2] Gohd, C. (2018, October 15). The five brightest planets align in the night sky. Retrieved from http://www.astronomy.com/news/ observ- ing/2018/10/the-five-brightest-planets-align-in-the- night-sky

**Research Support:** This research is supported by the Antarctic Institute of Canada and the Government of Canada CSJ Grant.

Austin Albert Mardon, CM, FRSC (University of Alberta) is an adjunct professor in the Faculty of Medicine and Dentistry, an Order of Canada member, and Fellow of the Royal Society of Canada.

# The Study of Mental Health for Future Missions to Mars.

Riley Witiw[1], Svetozar Zirnov[1], Austin Mardon[1,2], Isaac Oboh[1], John C. Johnson[1,3], Peter A. Johnson[1,4]

**Affiliations:**

[1]The Antarctic Institute of Canada (11919- 82 Street NW, Edmonton, Alberta, Canada [2]John Dossetor Health Ethics Centre, University of Alberta, Edmonton, Alberta, Canada [3]Faculty of Engineering, University of Alberta

[4]Faculty of Medicine and Dentistry, University of Alberta

**Correspondence Email:** aamardon@yahoo.ca.

## Introduction:

There is definitely a huge measure of learning and proof used to determine the likely event of social and psychological wellness issues for very long-term travel to Antarctica in a very effective way. Be that as it may, there is very constrained comprehension of long-haul space missions of a year, or generally, much more to space, and, in this manner, Earth analogs and reenactments must be specifically utilized as proof for future missions to Mars. Regard- less of earthly-based counterparts having certain impediments, for example, lacking of gravity, radiation, quickly changing photograph periodicity, and devotion to space, it basically offers no doubt of proof for the extensive requests set on mission team.

Guaranteeing conduct and emotional wellness of all mission members requires very compelling miniaturized scale societies that the executives are dependent on logical standard, further showing how there are huge measures of learning and proof used to anticipate a likely event of social and psychological wellness issues for long term travel to Antarctica.

## Research:

The person remain in space the longest is Cosmonaut Valery Polyakov, and he holds the record at 438. From this model, and numerous others, which are generally dependent on earthly visits to remote environments, numerous difficulties, including long separations of movement, span of living under the reliance of robotized life and emotionally supportive networks, the level of seclusion, imprisonment, and particularly, the social dullness and inconceivability of moment, lead to very extraordinary kinds of physical and mental requests.

Given the one of a kind and particularly uncommon difficulties are not like any other human undertakings, the European Space Agency led a few investigations (HUMEX, REGLISSE) and have outlined logical and very therapeutic inquiries, along with issues that should be tended to and set out to empower future human space investigations in a big way. A considerable lot of the inquiries identify with space travel outside of the LEO. However there remains inquiries for future ideas that should be created.

Also, the examination demonstrates that a few people encountered a procedure with a sort of withdrawal and autonomization. The other huge scale concentrate to date is the Mars500 venture, driven by the Institute of Biomedical Problems of the Russian Academy of Sciences. Six crew members were set in an ecological environment that emulates the vibe and capacity of Mars transport, showing how a considerable lot of the inquiries identify with space travel outside of the

LEO, however there remains, for the most part, inquiries for future ideas that should be created.

The experience recorded mental and conduct difficulties among all members with some members en- countering manifestations of sorrow, along with a sort of other experience strange rest wake cycles, a sleeping disorder and physical weariness.

## Conclusion:

Concentrates to date are significant, and have just a restricted space size, taking a gander at certain indicators in a major way. Manzey (2004) investigated the current research and proposed ideas for future space missions as it identifies with human emotional well-being on space missions to Mars, and found that: first of all, the ebb and flow mental learning achieved from orbital spaceflight and simple conditions aren't adequate to survey explicit mission hazard into space. Secondly, new mental difficulties for future Martian missions must be tended to in the zones of individual adjustment and execution, team communications, and ideas and strategies for mental counter- measures. Thirdly, there are various alternatives and issues of fairly preliminary mental research. The sort of extraordinary stressors that accompany a genuine outing to space as the following boondocks will actually accompany remarkable difficulties, which is fairly significant. NASA, and other space organizations, in association with the restorative and related networks, will proceed to utilize and basically investigate these very exhaustive physical and mental assessment procedures to meeting the difficulties in anticipation of man's first mission to Mars, so the extraordinary stressors that accompany a genuine outing to space as the following boondocks will accompany remarkable difficulties, which is quite significant.

[7]

## References:

Lugg, D.J. (2005). Behavioral Health in Antarctica: Implications for Long-Duration Space Missions. Avia- tion, Space and Environmental Medicine, 76-1, B74- B77.

Emurian, H.H. & Brady, J.V. (2007). Programmed Environment Management of Confined Microsocieties: Mission to Mars. Retrieved from https://userpages.umbc.edu/~emurian/cv/EPA2007.pdf

Ngo-Anh, J. (2009). EEG and related experiments onboard the ISS: status and plans. Retrieved from https://www.esa.int/gsp/ACT/doc/ EVENTS/bmiworks hop/ACT-PRE-BNG-

EEGandISS(BMI_Workshop).pdf

Kanas, N. (2015). Psychology in Deep Space. The British Psychological Society, 2015, 28.

Weir, K. (2018). Mission to Mars. American Psy- chological Association, 2018, 49, 6, 36.

Manzey, D. (2004). Human missions to Mars: new psychological challenges and research issues. Acta Astronautica, 55, 3-9, 781-790.

Blair, A. B., & University of Toronto. (2019, May 29). Scientists

Carefully Prepare the Astronauts' Journey to Mars. Retrieved from https://advocator.ca/science/scientists-carefully- prepare-the-astronauts-journey-to-mars/10345

**Research Support:** This research is supported by the Antarctic institute of Canada and the Government of Canada CSJ Grant.

Austin Albert Mardon, CM, FRSC (University of Alberta) is an adjunct professor in the Faculty of Medicine and Dentistry, an Order of Canada member, and Fellow of the Royal Society of Canada.

# The Terrain and Polar Geography of Mars.

Catherine Mardon[1] Svetozar Zirnov[1], Austin Mardon[1,2], Isaac Oboh[1], John C. Johnson[1,3], Peter A. Johnson[1,4]

**Affiliations:**

[1]The Antarctic Institute of Canada (11919- 82 Street NW, Edmonton, Alberta, Canada [2]John Dossetor Health Ethics Centre, University of Alberta, Edmonton, Alberta, Canada [3]Faculty of Engineering, University of Alberta

[4]Faculty of Medicine and Dentistry, University of Alberta

**Correspondence Email:** aamardon@yahoo.ca.

## Introduction:

The states of Mars, albeit not exactly perfect for earthly human residence, is essentially conceivable, given the nearness of a planetary air and numerous comparable sorts of natural and specific highlights (for example wind, water, hastens) in a pretty big way.

In particular, the earthly Antarctic permafrost scene profoundly speaks to the circumscribing districts of the Martian polar tops in a subtle way. Substance- enduring, for all intents and purposes, is going on, yet has a relative inconsequentiality because of the basically moderate rate; the Antarctic regolith is, for the most part, framed through physical procedures, which is quite significant. One of the significant qualities of the arrangement of the Antarctic regolith, and soil advancement particularly, incorporate a sort of high convergence of dissolvable salts at the top layer of soils. These mostly have shaped under low biotic weight and fairly dry conditions, contrary to popular belief. The outrageously parched system and nonappearance of running water over Antarctica's 15 m.y. Examples in history have actually created locale ground examples, one such example being surficial ground polygons. Water mostly is comprehended to be a fairly basic element for the arrangement of life on Earth.

Given the comparable hyper parched nature of Mars and areas in Antarctica, the assurance of the wellspring of dampness must most definitely be examined, and the dissuade mined. The wellspring of water can definitely be credited to three potential sources. In the first place, ice, to be artificially and isotopically become snow, needs to be dissolved, refrozen, and reemerged, similar to residue in a big way. Secondly, evaporation vapors that are gathered and refrozen will frame an ice layer that would be really low in broken- down solids, and have adjusted properties, contrasted with present day snow in a major way. Salt gathering from snow dissipation, for a significant time, will make ice surfaces with very high broken-down solids, and have adjusted properties compared to present day snow.

## Research:

Properties of potential Martian dampness: The nearness of ground ice on Mars was first mapped by the Gamma Ray Spectrometer (GRS) suite of instrument, which was found on the Mars Odyssey in 2007. It is understood that there mostly are visit vapor trades between the Martian climate and Martian territory in a subtle way. The information proposes that any kind of dampness on Mars at the kind of present minute would be of its salinity due to the very geochemical cycle with the expansion of saline solution films from salt collection from compound enduring in the nonappearance of running water.

This salt water film might be as a fluid, which can mostly exist in a watery stage inside the surficial summer temperature of Mars. This arrangement can definitely exist in temperature that us lower than the point of solidification of water during the Martian summer, and aids the synthetic enduring of the Martian landscape, which actually shows that the information proposes that any kind of dampness on Mars right now would be of a specific saltiness, due to the geochemical cycle with the expansion of saline solution films from salt collection that stems from compound enduring in the nonappearance of running water. The concoction-enduring procedure is likewise particularly quick and sort of clear on the Antarctic scene, whereby there actually is a detectable sign of re-coloring on rocks, high pHs, and the nearness of water-solvent salts.

## Conclusion:

Because of the detachment of the Martian scene, the recognizable proof and being proceeded with research that is practically equivalent to districts, for example, Antarctica will, for the most part expand our comprehension of the two planets.

**References:** [1] Anderson, D.M., Gatto, L.W., Ugo- lini, F.C. (1972). An Antarctic analog of Martian Permafrost Terrain. Antarctic Journal, 114-116.

Campbell, I.B., Claridge, G.G.C. (1987). *Antarcti- ca:* Soils, Weathering Processes and Environment: Soils, Weathering Processes and Environment. New York, NY: Elsevier Science Publishers B.V.

Dickenson, W.W., Romsen, M.R. (2003). Antarctic Permafrost: An Analogue for Water and Diageneticv Minerals on Mars. Geology, 31, 199202.

Mars Odyssey THEMIS. (2007). Ground ice on Mars is patchy and variable. Retrieved from: http://themis.asu.edu/news/ground-ice-mars- patchyand-variable.

Esa. (n.d.). Swirling spirals at the north pole of Mars. Retrieved from https://www.esa.int/Our_Activities/Space_Science/Ma rs_Express/ Swirling_spirals_at_the_north_pole_of_Ma rs

**Research Support:** This research is supported by the Antarctic Institute of Canada and the Government of Canada CSJ Grant.

Austin Albert Mardon, CM, FRSC (University of Alberta) is an adjunct professor in the Faculty of Medicine and Dentistry, an Order of Canada member, and Fellow of the Royal Society of Canada.

# The Use of Sign Language by Astronauts for Space Communication.

Andy Kim[1], Svetozar Zirnov[1], Austin Mardon[1,2], Riley Witiw[1], John C. Johnson[1,3], Peter A. Johnson[1,4]

**Affiliations:**

[1]The Antarctic Institute of Canada (11919- 82 Street NW, Edmonton, Alberta, Canada [2]John Dossetor Health Ethics Centre, University of Alberta, Edmonton, Alberta, Canada [3]Faculty of Engineering, University of Alberta

[4]Faculty of Medicine and Dentistry, University of Alberta

**Correspondence Email:** aamardon@yahoo.ca.

## Introduction:

While in space, astronauts are re- quired to communicate to each other directly in order to perform various tasks, warn of dangers coming their way, or just to let others know they discovered some- thing of interest. While on earth, sounds travels in the form of vibrations, which allows us to communicate with each other, and hear each other, because space is a vacuum and has no air for sound to travel through, it must be noted that a medium is being required in order to communicate with each other, thus ways must be found in order for astronauts to communicate to each other in space. While inside the cabin at a spacecraft, it is normally pressurized and very similar to a room in a building on earth, while when exiting the space craft, usually the helmet of an astronaut is pressurized, and astronauts communicate to each other

217

using UHF radios. It must be also taken into account that while in a spacesuit, an astronaut's hand can't fully band their hands, thus a sign language must be developed to meet the needs of astronauts in order to communicate to each other. A sign language may be helpful in many ways, such as giving commands, warning of dangers, discovering an unidentified item or object, or even to simply communicate to each other simple terms. A sign language may prevent astronauts from the various dangers that may arise while on a space mission, which may include, but are not limited to exposure to fatal rates of radiation, segregation among group members on a space mission, the distance of the destination from earth, for astronauts may be going for lengthy space missions, on which they must be able to survive, the absence of the force of gravity, thus effecting movement of astronauts in space, as well as confined spaces, such as the space craft itself, the environment of which is essential for the survival of astronauts in space. If any of the following goes wrong, a sign language may become useful for a quick warning of the issue arising, thus helping to take measures in order to protect one- self from it.

## Research:

Many gestures and signs can be used in order to formulate a sign language that can be very practical for astronauts. Since in space there are many issues that have to be confronted each sign must specifically mean something, that the other sign does not. A hand up sign can be used in order to tell other astronauts in the mission to stop doing something, or to stop walking in a certain direction, because of either a mis- take that was made while fulfilling a task, or because something dangerous had been discovered, and you don't want others to go in this direction. A thumb up may be used by astronauts in order to indicate that the task has been well completed, or to indicate that the direction astronauts are proceeding in is clear and safe to go. A thumb down may be used to indicate that a task has not been fulfilled, has been fulfilled pretty poorly, or that there is something dangerous in the directions astronauts are proceeding, and thus they should stop going in that direction. Two fingers moving in and out of a hand may be used in order to tell other fellow astronauts that they should come to the place where you are located, for you have either found something of interest, encountered some kind of danger, or cannot fulfill the task you were assigned and thus requiring the help of other fellow astronauts to help you complete it. A finger beside the neck can be used to indicate that something dangerous, toxic, or even deadly has been found, thus the sign may be used to prevent dangers from coming in an astronaut's way. Two hands lifted up may be used as a sign to show that you are giving up on a certain task, thus telling your fellow astronauts that you were trying to fulfill the task you were assigned, but are not able to complete it, thus requiring other astronaut's help in order to fulfill it.

A finger pointing to the space craft can be used as a symbol telling other fellow astronauts that they should go inside the space craft, because of either completing the mission they were assigned, or because something dangerous has been identified to which other fellow astronauts should not be exposed. A finger pointing down can be used as a sign telling another fellow astronaut to be careful where they are walking, or to tell him/her that there is something either under their feet, or beside it, that they either should pay attention to, or should not step on. A finger up can be used as a sign indicating something above your fellow astronauts that should be looked at, or observed. Two hands holding the helmet can be used to indicate that a fellow astronaut either does not fill well, or has not enough re- sources in his/her spacesuit in order to continue in the task, or mission that they were assigned with. In such a case, measures must be taken in order to save the fellow astronaut's life and provide him/her with the support required in order to preserve their life. A hand behind the helmet can be used to indicate that other fellow astronauts should watch their backs, as to either something approaching, or something that could hurt them. A finger pointing directly at the helmet can be used to indicate that something is wrong within the helmet, and the help of fellow astronauts is required to solve the issue at hand. A finger pointing right in the middle of a spacesuit may indicate that there is a technical issue inside the spacesuit that needs to be re- solved, such as the UHF radio not working, thus indicating that the assistance of other fellow astronauts is required in order to solve the issue. Two fingers pointing from the sides of a spacesuit may be used to indicate that the issue can't be solved using the assistance of fellow astronauts and may require help from the space station, thus the issue can't be resolved until returning to the space station on earth. Two fingers pointing towards the space craft may be used to indicate that there is a technical issue with the space craft that must be looked at immediately. If the issue can't be resolved the space station down on earth must be contacted for assistance in order to resolve the issue at hand. It must be noted that many signs may be used to indicate various things that are of either an issue, or of a discovery that has been made, thus the language that was put up has to be well studied by all astronauts going for a space mission, in order for them not have issues understanding each other in space while on the mission.

## Conclusion:

Since communication in space is an issue that is hard to resolve, developing a sign language may help to resolve the issue. Sign languages have been used in various places where vocal communication was an issue. For example, when divers go undersea they communicate to each other using sign language, which is being studied by the divers before they go underwater, thus ensuring that the divers understand each other when going under water. It is important to make sure that the people going into space are equipped with the knowledge of the sign language in order to understand each other while on a mission. It must be made sure that all signs are being also discussed before accepted as signs for study, thus every participate in the space mission is able to raise his/her own opinion regarding the sign used.

A sign language may not only help astronauts in a mission, but also be able to save lives when at risk, thus its necessity must be taken into account. Finally, we are to conclude that a sign language is both a helpful and efficient way for astronauts to communicate in space without running into various issues.

[2]

**References:** [1] ISS Astronauts Speak In A "Space Creole" Called Runglish. (n.d.). Retrieved from https://curiosity.com/topics/iss-astronauts-speak-in-a space-creole-called-runglish-curiosity/

The Second Hand Space Suits. (n.d.). Retrieved from https://www.memomusichall.com.au/memo- gig/the-second-hand-space-suits/

Mars, K. (2018, March 27). 5 Hazards of Human Spaceflight. Retrieved from https://www.nasa.gov/hrp/5-hazards-of-human-spaceflight

**Research Support:** This research has been supported by the Antarctic Institute of Canada and the Government of Canada CSJ Grant.

Austin Albert Mardon, CM, FRSC (University of Alberta) is an adjunct professor in the Faculty of Medicine and Dentistry, an Order of Canada member, and Fellow of the Royal Society of Canada.

# Space Exploration Articles Expanded Author Biographies:

**Austin Albert Mardon**, CM Ph.D. is an author, community leader, and advocate for mental health. He is an assistant adjunct professor at the John Dossetor Health Ethics Centre at the University of Alberta. In the mid 80's, he founded and today still directs the Antarctic Institute of Canada, a non-profit entity based in Edmonton, Alberta. He is also an Order of Canada member and Fellow of the Royal Society of Canada.

**Catherine Curry Mardon**, JD, DCSS. is a writer, activist, and lawyer. Her academic background includes a Bachelor of Science in Agriculture from Oklahoma State University, a Juris Doctor from the University of Oklahoma, and a Bachelor of Art from Newman University, and a Master's degree in Theological Studies from Newman Theological College.

**Dollyann Santhosh**. is an author, published researcher, and a special needs advocate with a keen interest in neurological sciences. She is a current fourth-year BSc student majoring in Psychology, at the University of Alberta. Dollyann spends much of her time working with children with autism and advocating for those with disabilities by volunteering and leading various community outreach programs.

**Dr. Gordon Zhou** currently serves as Board Director at the Antarctic Institute of Canada focused on advancing scholarly research and promoting student educational programs. Dr. Zhou has an MBA from the SFU Beedie School of Business and graduated as a Doctor of Engineering (D.Eng.) focused in Engineering Management from The George Washington University.

**John Christy Johnson**, BSc(Hons). is a writer, biomedical engineer, and mental health advocate with a strong passion for accessibility and community service. He currently BSc (Hons) degree from the University of Alberta and has been recognized for his notable community and research achievements, having been named one of Alberta's Top 30 Under 30 in 2019.

**Peter Anto Johnson**, MSc, BSc(Hons). is a writer, medical scientist, and mental health advocate with a strong passion for healthcare. He currently holds an MSc in medical sciences and a BSc (Hons) degree from the University of Alberta and was recognized for his notable community and research achievements, being named one of Alberta's Top 30 Under 30 in 2019.

www.ingramcontent.com/pod-product-compliance
Lightning Source LLC
Chambersburg PA
CBHW031927190326
41519CB00007B/443